THE AGE OF BIFURCATION

THE WORLD FUTURES GENERAL EVOLUTION STUDIES

A series edited by Ervin Laszlo
The General Evolution Research Group
The Vienna International Academy, Austria

This book is part of a series. The publisher will accept continuation orders which may be cancelled at any time and which provide for automatic billing and shipping of each title in the series upon publication. Please write for details.

THE AGE OF BIFURCATION

Understanding the Changing World

Ervin Laszlo
The Vienna International Academy
Austria

Foreword by Ilya Prigogine

Gordon and Breach
Philadelphia Reading Paris Montreux Tokyo Melbourne

Gordon and Breach Science Publishers

5301 Tacony Street, Drawer 330
Philadelphia, Pennsylvania 19137
United States of America

Post Office Box 90
Reading, Berkshire RG1 8JL
United Kingdom

58, rue Lhomond
75005 Paris
France

Post Office Box 161
1820 Montreux 2
Switzerland

3-14-9, Okubo
Shinjuku-ku, Tokyo 169
Japan

Private Bag 8
Camberwell, Victoria 3124
Australia

Library of Congress Cataloging-in-Publication Data

Laszlo, Ervin, 1932-
The age of bifurcation: understanding the changing world / Ervin Laszlo ; foreword by Ilya Prigogine.
p. cm. — (The World futures general evolution studies ; v. 3)
Includes bibliographical references and index.
ISBN 2-88124-491-2 (hardcover)
1. Science—Philosophy. 2. Science—Social aspects. 3. Chaotic behavior in systems. 4. Bifurcation theory. I. Title. II. Series.
Q175.L285 1991
501—dc20 91-23915
CIP

Printed in the United States of America.

CONTENTS

INTRODUCTION TO THE SERIES

The World Futures General Evolution Studies series is associated with the journal *World Futures: The Journal of General Evolution.* It provides a venue for monographs and multiauthored book-length works that fall within the scope of the journal. The common focus is the emerging field of general evolutionary theory. Such works, either empirical or practical, deal with the evolutionary perspective innate in the change from the contemporary world to its forseeable future.

The examination of contemporary world issues benefits from the systematic exploration of the evolutionary perspective. This especially happens when empirical and practical approaches are combined in the effort.

The World Futures General Evolution Studies series and journal are the only internationally published forums dedicated to the general evolution paradigms. The series is also the first to publish book-length treatments in this area.

The editor hopes that the readership will expand across disciplines where scholars from new fields will contribute books which propose general evolution theory in novel contexts.

FOREWORD

It is a pleasure to write the foreword to Ervin Laszlo's latest book, *The Age of Bifurcation*. This is not an easy task, as his book deals with the complex problems facing our societies in the future. Still, I am happy to undertake this task as I feel that Laszlo's study represents a remarkable coincidence: at this very moment, mankind is living through a crucial time of transformation while science is undergoing a spectacular transition. More and more, an ever-increasing number of scientists perceive that a new paradigm is taking shape. Everywhere we see fluctuations, evolution, diversification. This is true not only on the level of macroscopic phenomena, as in chemistry, but also on the microscopic level in particle physics and on the vast scale of modern cosmology.

Ideas about the role of time in natural science have changed radically. One may recall the Heisenberg paradigm expressed in quantum uncertainty relations. This was indeed a fundamental step: quantum mechanics was essentially the first science to be forced to give up the deterministic description. The process was difficult. Einstein, for example, was opposed to the statistical character of quantum mechanics; in one of his last papers he wrote that uncertainty applied only to the microscopic problems described in quantum mechanics; in the macroscopic world, he was certain determinism continued to be the rule. On this point the developments of the last few decades do not support Einstein's belief.

We now know that for systems in far from equilibrium conditions there are some behaviors that can only be described in terms of a mixture of determinism and probabilities. This is true even at the macroscopic level. A well-known example is the so-called Bénard instability, due to a vertical temperature gradient being set up in a horizontal layer of liquid. As heat begins to be conveyed through convection, for some threshold value of this gradient (meaning at some rate of heat flow), the fluid's state of rest becomes unstable; this corresponds to the coherent motion of huge assemblies of molecules. This convection is a symmetry-breaking phenomenon: two points which were "equivalent," being "almost" at the same place before the initiation of the heat constraint, are now involved in distinct convection rolls. We must even say that for a given region of space, the choice between the various possibilities depends on statistical fluctuations, described by probabilistic laws.

This is a general feature of unstable dynamic systems. According to our present view, whatever the (finite) precision that we possess regarding the initial conditions, we can only predict the probability that one of the many possible structures will come about. This is not in any way a defeat for the human mind. Far from that. In the appendix entitled "Basic Concepts of Evolutionary Systems Theory," Laszlo shows how the new scientific conceptions suggest that "the selection from among the set of . . . alternative steady states is not predetermined."

The new recognition of the philosophical and practical implications of our human, finite predicament prompts us to give up the traditional ideal of omniscience. Indeed, that dream of classical physics had theological roots. This becomes clear in even a cursory reading of the correspondence between Clarke (who spoke for Newton) and Leibniz. Leibniz begins the argument by saying that Newton seems to have a poor idea of God since for him the Creator would be obliged to retune his work from time to time, thus being a bad watchmaker. For Leibniz it was inconceivable that God would not know at any given moment what would happen in any arbitrarily distant future. In the same tradition, a major current of modern physics tried to negate the importance, and even the existence, of time. Einstein himself wrote that time is an illusion, and through this belief, he thought, scientists would come closer to God. From this perspective, the very proof of valid knowledge turns out to derive from the conceptual elimination of time. This theological prejudice is now crumbling. Together with the ideal of omniscient science, also on its way out is the dualistic description of an almighty humanity, subordinating nature to itself.

The title of this book, *The Age of Bifurcation*, is well chosen because with the concept of bifurcation the historical category of "event" enters contemporary science. An event is something that cannot be deterministically predicted. The position of the Earth around the Sun in a given number of years could hardly be considered an event, while obviously the birth of Mozart was an event in the history of Western music.

Laszlo cites the cyclical views of Spengler and Toynbee on the fate of civilizations that grow and die and are born again. However, he advocates instead the emergence of a new civilization, more adapted to the evolving planetary situation. I hope that he is right: as he points out, the epoch in which we live is crucial.

I have always been struck by what we might call a kind of synchronicity in history: great innovations such as pottery or plant cultivation seem to have appeared more or less at the same time all over the world. But this synchronicity was broken in the nineteenth century with the advent of cultural destabilization, of inequality, of an epoch when humanity divided into savages and civilized people. In the present day, we try to go beyond this dichotomy to attain a more universal vision of human dignity.

In conclusion, let me express a somewhat naive view. I was always under the impression that we paid dearly in the transition from the tribal stage to the building of the modern state (e.g., slavery, domination, large-scale wars). In the Western world there is a tendency to idealize the prehistorical or "natural" condition. Think of Rousseau, Bachofen or Lévi-Strauss. For the latter, anthropology is "entropology"—the history of cultural decline, of the loss of genuine values. But we now have hope that with our achievements, both theoretical and experimental, with our immensely improved capacity of producing wealth, and with our new facilities for interpersonal communication, we can come at last to a form of civilization where an increasing number of people have the possibility to manifest the creativity which, I believe, is present in every human being. Is this the beginning of such a new age? We are still too heavily involved in the planetary transformation underway to reach a firm evaluation, but perhaps—and this is my hope—succeeding generations will see our time as the beginning of a great age of bifurcation—and will look upon this book as the herald to that age.

Ilya Prigogine

PREFACE
Interesting Times

There is an old Chinese curse that says, "May you live in interesting times." I don't believe anyone would dispute that we indeed live in interesting times—most interesting. But whether these times are accursed or blessed is probably less clear to many people. My own orientation is that neither is the case: we are under no dark cloud, nor do the heavens smile upon us. What the world will be like for us and our progeny is very much up to us.

This belief constitutes the basic underlying premise upon which our work at The Vienna International Academy is predicated. This century has seen the advent of an era in which the range of possibilities for life—for the quality of life, and perhaps even for the persistence of life—on our planet is very much, if not entirely, to be determined by what we, the human inhabitants of the planet, do. And what we do, we assert, will be a direct consequence of what we believe and know—of how we approach problems and situations.

These same words may have been said (and no doubt were) countless times before. But the words do not mean the same today as they did in the past. In previous eras, these words were meant strictly in the local sense. What people did may have profoundly affected their immediate environment—their health and longevity; their family and relationships; their community; perhaps even their city or country. It was extraordinary for any individual or even any group of people to effect large-scale changes on the planet. It was virtually impossible to affect the Earth as a whole because the medium of interaction, as it were, was so "viscous." Information travelled slowly; natural resources yielded wealth only with great effort and over long periods of time; new knowledge was slow in coming; human impact on the global environment was on the scale of, at most, minor perturbations. The world changed slowly.

Our day has seen developments which lift the underlying notion—that what the world will be like in the future depends on what we humans believe, know, say and do—out of the realm of platitude and into the realm of scientific principle. This book is, in the first instance, an argument for accepting this premise with a new sense of seriousness. It is also a précis for adopting a particular orientation to global problems—an orientation we call the general evolutionary systems approach—with a new sense of vision. Finally, it is a call for the application of this approach to the critical problems that we as a species, and the planet as an inhabitable home, face with a new sense of purpose.

The issues are burning, the stakes enormous, and the options impressive. These truly are interesting times. This stands to reason: we live in humankind's greatest age of bifurcation.

Ervin Laszlo

INTRODUCTION
The Challenge

1. Global Problems

The problems that beset humanity in the 1990s are not only political, but social, economic and environmental as well; they are closely interconnected, of worldwide scope and universal impact. The problems arise because human societies, having grown and developed throughout the course of history, have reached planetary dimensions. The historical process has extended the size and dominion of societies from the local to the national level, from the regional to the global level. It has shifted people from life in basic kinship structures with low-energy technologies to the complex structures and high-energy processes in which people now conduct their existence. The vast acceleration of the historical growth and development process in the twentieth century has created social rifts and polarizations, resulted in highly differentiated rates of economic development, and has had runaway impacts on nature.

2. The Impacts of Technology

Guiding this historical development process have been innovations in technology: the continual extension of the power of human muscle, of human sense organs and the human brain through artificial systems and machines. The first industrial revolution mainly extended the power of human muscle in the manufacturing and transportation of goods. A parallel extension of human sense organs via telescopes, microscopes and other scientific and medical instruments followed, and led to major discoveries in the far reaches of cosmic space, the depths of the atomic world, and in the human body itself. In turn, the technological innovations of the latest industrial revolution extended the power of the human brain by creating breakthroughs in the gathering, storing and processing of information.

Every technological revolution changed and reformed the structures of the societies and organizations that employed these innovations. In the late twentieth century, through the use of an elaborate communication network, humanity is being propelled from a nationally-based industrial society to a globally interconnected socioeconomic system.

3. The Rhythms of Change

The historical development of societies, while progressive, has been neither continuous nor smooth. Transitory epochs of instability and chaos have punctuated extended periods of stability and order. The epochs of comparative disorder have been overtures to profound changes and transformations; they exemplify the "creative chaos" discovered in the computer modelling of complex dynamical systems. The current transition from a national industrial to a global information society is subject to a recurring series of instabilities. Today, these instabilities are transforming Soviet and Eastern European political systems from dictatorial socialism to free-market democracy, even as they triggered the process whereby former colonial systems did not join the international community as equal social and economic partners, but entered the state of underdevelopment which characterizes much of the Third World.

4. The Emerging Threat

Unless timely action is taken, another wave of instability could beset the structure of contemporary society: the wave triggered by increasing levels of stress and degradation in the biosphere. The global ecology, the matrix for all human and industrial natural resources, is a dynamic system in its own right. Human and technological impacts on it produce irreversible changes, including different, and conceivably less favorable, thermal and chemical equilibria. The present generation can neither return to past conditions nor extend ecological impacts in a linear fashion. The instability triggered by these changes could become the ultimate arbiter of the success of the current transition to a globally extended information-based society.

5. The Remaining Option

System stress is not a new phenomenon. Beginning with the Neolithic revolution some ten thousand years ago, people disregarded their environment as a closed ecological system, and as populations grew and demands on the environment increased, productive lands became overstressed. At this point, people either became more efficient in the use of their natural resources or moved to less exploited environments. In the late twentieth century, humanity is again reaching the limits of its natural resource base, but of the two classical solutions mentioned above, one is no longer an option: on this planet hardly any unexploited lands or unimpaired environments are left. Only the better exploitation of the existing resources remains a viable option, making efficient and responsible use of the technologies that drive the historical development process.

6. The Challenge of Self-Regulation

The challenge to our generation is to develop the modes of thinking, valuing and living capable of regulating the global socioeconomic and ecologic system before it becomes overstressed. Unlike regulatory mechanisms in nature which are genetically coded and automatically activated whenever thresholds of stability are overstepped, the regulatory mechanisms of human society are culturally dependent on the values and insights of living generations. Culture-coded mechanisms evolve more rapidly than genetically coded ones, but they also become more rapidly outdated. The technological revolution of the last few decades has outdistanced established views and beliefs—in the 1990s obsolescence in regard to the values and practices by which contemporary societies cope with each other and with their natural environment has become a genuine threat to human well-being and survival. A timely and purposive regulation of the global socioecologic system is possible: the knowledge-base exists, the technologies for applying it are there, and a new consciousness to support it is evolving in the mind of the general population. The development of scientific knowledge concerning the dynamics of evolutionary processes and the technological application of that knowledge provide an effective platform for timely self-regulation, inspiring up-to-date thinking and effective, responsible decision-making.

7. The Need for Evolutionary Competence

The relevance and effectiveness of this new knowledge applies to all spheres of action in the public as well as in the private sector. Corporations such as IBM, Nissan, Asea-Brown-Boveri and others who knew how to grow from the national to the international and then to the global level became industrial leaders. Governmental leaders such as Mikhail Gorbachev, who launched an evolutionary process, and Boris Yeltsin, who knew how to stay on top of it, became leaders of their society. But corporations can fail and leaders can fall; for assured success the intuitive feel of the nature of evolutionary processes needs to be replaced by reliable competence in coping with their dynamics. Evolutionary knowledge confers unique benefits on the individuals who possess it and the societies in which those individuals live and work. To acquire such competence is a challenge confronting all segments of society, ordinary citizens as well as leaders in government and business.

PART ONE

THE PATH TO THE PRECIPICE

CHAPTER 1

Birth of a Word; Birth of a Science; Birth of an Age

In years to come (if it has not happened already), people will find it difficult to believe there was once a time when educated people, even scientists, were ignorant of the scientific import of the word "bifurcation." The word has certainly existed for a long time as part of ordinary language—just as words like inertia, cell, interval and attractor existed long before Galileo, Schwann, Lorentz and Ulam, respectively, gave these words new scientific meanings. Outside the context of the science advanced by these individuals (kinematics, microbiology, relativity, and chaos theory), these words had mundane explanatory powers that paled next to their scientific meanings.*

In a sense, these words were born anew, just as assuredly as those scientific terms that were freshly coined—words like vector, proton, lysosome and quark. With the birth of these words were born new sciences—new ways of looking at things, of investigating the world, and of manipulating nature—that rendered all previous work in these fields obsolete, irrelevant and (sometimes unfairly judged to be) misconceived. In certain instances, the new sciences that sprang up around these words have had so great an impact on our lives as to give rise to a new age, a world qualitatively different from that of the earlier period.

---

*Consider how different our attitude will be toward the term "string" should String Theory turn out to be the primary descriptive and predictive formulation of particle physics. For the time being, however, the power of the word is equivocal; the jury is still out on String Theory. Such is not the case with bifurcation and the other terms of Chaos Theory.

Such was certainly the case in the transition from pre-Euclidean to post-Euclidean times, and the same may be said for the new ages spawned by Newton, Darwin, Freud and Einstein. New words frame new concepts that form the superstructure of whole new sciences that give birth to new epochs for humankind.

Bifurcation is just such a term, lying at the core of a science that offers a means of understanding systems and phenomena previously beyond the grasp of any science. In fact, of all the terms that form the lingua franca of chaos theory and the general theory of systems, bifurcation may turn out to be the most important, first because it aptly describes the single most important kind of experience shared by nearly all people in today's world, and second because it accurately describes the single most decisive event shaping the future of contemporary societies. Yet, except for a few researchers on the cutting edge, few know what the word means and how to apply it. Even the 1985 edition of the *Encyclopedia Britannica* contains virtually nothing on either bifurcation or chaos theory.

Just what *is* bifurcation? Like chaos, this is a word that means something other than it used to. Chaos used to mean disorder and confusion; it now means subtle, complex, ultrasensitive kinds of order. Bifurcation in turn used to mean splitting into two forks (from the latin *bi*, meaning two, and *furca*, meaning fork). But today bifurcation means something more specific than that: in contemporary scientific usage this term signifies a fundamental characteristic in the behavior of complex systems when exposed to high constraint and stress. It is important to know about this meaning because we ourselves, no less than the societies and environments in which we live, are complex systems exposed to constraints and stress. In fact, in many contemporary societies levels of constraint and stress are now reaching critical dimensions.

An acquaintance with the new meaning of bifurcation belongs to the essential knowledge of our age. This meaning is specified in some of the newest and most advanced branches of the natural and mathematical sciences. The relevant sciences include nonequilibrium thermodynamics (also known as the thermodynamics of irreversible processes), and dynamical systems theory (the latest offshoot of classical dynamics). However, there is no need to be alarmed: notwithstanding its technical origins, the new meaning of bifurcation is not difficult to grasp.

In nonequilibrium thermodynamics—the natural science that deals with the dynamics and evolution of systems in the physical universe—bifurcation refers to the behavior of complex systems in states and conditions that are far from equilibrium. Bifurcation occurs when such systems are destabilized in their environments, stressed out of states in which they could comfortably remain virtually forever. Because complex systems in the real world are nearly always far from "equilibrium" (which in this context does not mean balance and rest but a dynamic state where internal forces keep a system from lapsing into randomness), changes can frequently occur that upset the balance between the internal forces structuring the systems and the external forces that make up their environment. When that happens, sudden and non-linear "chaotic" processes take place that either restructure the system and propel it along a trajectory that becomes more and more complex, leading ultimately to the evolution of life—and perhaps also of mind and consciousness—or else they fatally perturb the system and cause its disintegration. In the science of nonequilibrium thermodynamics the evolution of complex systems is always irreversible because the only alternatives available to the system are those of increasing complexity, or else total extinction. Thus, systems described in the science of nonequilibrium thermodynamics display a definite direction of temporal development, or a "Time's Arrow," contrary to that in classical thermodynamics.

Bifurcation has a more abstract but no less relevant meaning in Dynamical Systems Theory, the science that gave birth to the new concept of chaos as a complex and unpredictable form of order. In dynamical systems theory a system is described in a "phase space" graph of the total possible states the system can be in. The system can respond to certain "attractors," or forces that cause the system to develop along certain "trajectories" (also called "time-sequences") in phase space. Since these attractors act on the system as a whole and cause it to change dynamical qualities beyond (but not excluding) position, attractors are not forces in the classical sense. (Another way of putting it: classical forces in which a field—such as a gravitational field or an elastic collision—prescribes the orbit or locus of a mass is the most primitive and simplest example of "an attractor defining a system's trajectory in phase space.")

When a system is "stressed" beyond certain threshold limits (for example, it is heated up, or its pressure is increased), it shifts from one set of attractors to another and then behaves differently—or, to use the language of the theory, the system "settles into a new dynamic regime." It is at the point of transition that a bifurcation takes place. The system no longer follows the trajectory of its initial attractors, but responds to new attractors that make the system appear to be behaving randomly. However, the system is not behaving randomly, it is merely responding to a new set of attractors that give it a more complex trajectory.

The term bifurcation, in its most significant sense, refers to the transition of a system from the dynamic regime of one set of attractors, generally more stable and simpler ones, to the dynamic regime of a set of more complex and "chaotic" attractors. Mathematicians use computer modeling and simulations to study various kinds of bifurcations, categorizing them in terms of the bifurcation itself as well as of the dynamic regimes into which they lead. Bifurcations are called "subtle" if the transition is smooth and continuous; "catastrophic" if the transition is abrupt and the result of mounting attractor stress; and "explosive" if it is the result of sudden and discontinuous factors that wrench the system out of one regime and into another.

Once in the new dynamic regime, the system can act in many ways. It may respond to new attractors that impose a new order on the system, keeping it in a state of fluctuation between discrete values in the regime (known as a Turing bifurcation), or the system may fluctuate wildly among many values, failing to settle on any one or set of values (in which case, we are dealing with a Hopf bifurcation). Finally, the bifurcation may be simply a transitory stage by which the system passes through a regime in order to find a new area of stability, in which case the bifurcation is a "window" to a stable dynamic regime for the system.

When we consider the phase space description of a system, we can see the aptness of the term bifurcation. The system proceeds in its stable state along well-formulated trajectories, until one parameter exceeds a threshold limit. At that point, the trajectory forks and the system enters a region of phase space where it behaves differently and assumes new and different values. It follows another trajectory, dancing to the tune of new attractors. It is important, however, that in the course of their evolution, complex nonequilibrium systems describe a trajectory in their state space marked by a definite pattern. When bifurcation occurs, the fact that we cannot predict the exact trajectory it will take does not prevent us from seeing and predicting basic patterns that the evolving system will display over time.

The bifurcation process tells us that when a system is pushed beyond its threshold of stability, it enters a phase of chaos. This is not necessarily fatal to the system; it could also be a prelude to a new development. In viable systems chaos gives way to higher forms of order. However, the relation between pre-crisis and post-crisis order is never linear—it is not one of simple cause and effect. Through the process of bifurcation the evolution of nonequilibrium systems is saltatory and nonlinear. As a result, bifurcation is full of surprises.

In the realms of nature it is impossible to predict which way a bifurcation will cut. The outcome of a bifurcation is determined neither by the past history of the systems nor by their environment, but only by the interplay of more or less random fluctuations in the chaos of critically destabilized systems. One or another of the fluctuations that rock such systems will suddenly "nucleate." The nucleating fluctuation will amplify with great rapidity and spread to the rest of the system. In a surprisingly short time, it dominates the system's dynamics. The new order that is then born from the womb of chaos reflects the structural and functional characteristics of the nucleated fluctuation.

These seemingly abstract findings are highly relevant in the concrete context of today's social change. The social, economic, political systems in which we live are complex and unstable; sooner or later their evolutionary paths must bifurcate. Our world, no less than the world of nature, is subject to phase changes. Bifurcations are more visible, more frequent, and more dramatic when the systems that exhibit them are close to their thresholds of stability—when they "live dangerously." This is just the behavior our complex societies are exhibiting in the late twentieth century.

Fortunately, in society, bifurcation is not necessarily the plaything of chance. After all, the actors that create the crucial fluctuations are conscious human beings. If they come to know the nature of the process in which they act they can steer it. They can bias the otherwise random interplay of fluctuations "from the inside." They can create new lifestyles, alternative patterns of behavior, appropriate technological innovations, and environmentally conscious and effective social and political movements. When established beliefs and practices become nonfunctional and obsolete, the search for more functional and effective ideas and behaviors gets under way. Many new concepts and strategies surface, and some of them could catch on. Replicated and disseminated through the rapid communication networks of our age, these "hopeful monsters"—as viable mutants are sometimes called in biology—can become the major factors shaping the future.

Historical examples of societal bifurcation are legion, especially in the twentieth century. Tsarist Russia was driven beyond its threshold of stability by internal dissension and a lost war in 1917. The system broke down, and out of the chaos of the "October Revolution" emerged Lenin and the unexpected Marxist regime of the Bolsheviks—the same regime that broke down in 1991 in another unexpected bifurcation. The Weimar Republic in Germany reached its own threshold of stability in the late 1920s, and the chaos of a bankrupt and discontented society gave birth to the monstrosity of Hitler and national socialism. In 1948, China's nationalists faced a crisis in the throes of which Chiang Kai-Shek and the dominant Kuomintang regime collapsed and Mao and his ragged band of a thousand came to power. In more recent years, similar instabilities have rocked Cuba, Nicaragua, Ethiopia, Angola, Iran, and the Philippines—among many others. Unexpected outcomes were the rule rather than the exception.

The instabilities themselves can have diverse origins. They can be due to the effect of insufficiently assimilated or badly applied technological innovations—they are then instances of what I have called "T-bifurcations." They can be triggered by external military conquests or by internal social and political conflicts, forming "C-bifurcations." Or they can be caused by the collapse of the local economic-social order under the impact of mushrooming crises, giving rise to "E-bifurcations."* Whatever their origin, the instabilities are likely to spread to all sectors and all segments of society. They then open the door to rapid and fundamental change.

Although conflict-generated C-bifurcations sometimes occur (and receive the most media attention), the great majority of the bifurcations that began to rock societies in the postwar years—and are likely to continue to do so throughout the remaining years of this century—are a combination of T- and E-bifurcations. They are the result of the opening of underdeveloped social and economic systems as they are suddenly exposed to the full impact of global flows of information, technology, trade, and people. When politically isolated or semi-isolated systems open up, they are caught up in the vortex of the globalizing modern world. Their people want the fruits of "modernization" but their work habits, values, behaviors, and above all their institutions and their production, distribution and consumption patterns are not able to cope. The global flows serve at the most a small elite group, often no more than a few percent of society. A small number of privileged people are quickly "Westernized" and "modernized" while the rest remain underdevel-

---

*For a more detailed description of such bifurcations, see Ervin Laszlo, *Evolution: The Grand Synthesis*. Boston and London: Shambhala New Science Library, 1987, Chap. 9.

oped and become increasingly frustrated. As long as the political system is stable and its leadership authoritarian, repression and dissimulation give a superficial semblance of stability. But as soon as the dictatorship breaks down, the situation becomes explosive. Frustration fuels reforms that shade into revolt. Society turns chaotic, and its behavior becomes unpredictable.

In the last forty years, there have been two great waves of such global flow-triggered bifurcations. Both were acclaimed as humanistic and long overdue reforms, and both proceeded from praiseworthy motivations. Yet both encountered unexpected problems and had entirely unforeseen consequences. The first of these waves unfolded under the aegis of "decolonization," and the second under that of "glasnost."

Decolonization opened previously semi-isolated traditional societies to the modern world. Colonial peoples were connected with unsymmetrical ties to their colonial masters; they were carefully detached from the rest of the world. The "mother countries" did not want their colonies to access technologies and information that would pave the way to self-sufficiency and fuel demands for independence. Decolonization opened the floodgates. Global flows of information, technology, trade, and people caught the unprepared newly liberated people in a disorienting and disrupting vortex. With but few exceptions, they fragmented and polarized, and were unable to set society and economy along the path of socioeconomic development. Expensive national airlines, flashy limousines and tourist facilities, and a few elite hospitals and schools contrasted painfully with quagmired villages and empoverished rural populations. Foreign powers and multinational corporations exploited the situation for their own purposes, bringing benefits only to the already privileged sector that controlled the strings and entered the markets. In vast regions of the Third World the process of development broke down, poverty became endemic, and national debt piled up. This was, and remains, the situation in many parts of sub-Saharan Africa, Central and South America, the Caribbean, and Western and southern Asia. If the people of many of these lands have not staged a full-scale uprising it is only because deprivation and polarization have not yet reached the crucial threshold—have not yet reached the point of bifurcation. But reaching it seems only a question of time.

The second wave of T- and E-bifurcations came with Mikhail Gorbachev's policy of glasnost. Glasnost opened the Second world of socialist countries to the First-world generated flows of information, technology, trade, and people, much as decolonization opened the Third world. The outcome was just as unexpected, while the rebellions were quicker to emerge. As soon as the people of Eastern and Central Europe were freed from the constraints of autocratic one-party rule, they organized themselves, demonstrated, and rebelled. Poland and Hungary led the way, East Germany, Czechoslovakia, Romania, and the Baltic states soon followed. The rest of the former Soviet empire soon followed.

Throughout the Soviet Union and Eastern Europe, glasnost was to produce perestroika—"openness" was to lead to "restructuring." But in Eastern Europe, perestroika transformed into revolt and in the USSR it stagnated even before it had properly begun. The full overhauling of a communist economy, it now becomes apparent, entails major adjustments that amount to a complete replacement of the socialist system of production and the one-party system that administered it. This is what happened in Poland, Hungary, Romania and Czechoslovakia, where multi-party pluralism is the order of the day. Yet, even in these countries the economy is on the verge of breakdown, with skyrocketing foreign debt and the remnants of an inefficient, cumbersome and overinflated public sector that devours resources and gives little in return. Wherever it reaches, instead of perestroika, glasnost produces bifurcation: not the restructuring of the communist system, but the onset of a period of chaos, fluctuation, and uncertainty. The crumbling of the Berlin wall was a fitting symbol of the process: as the wall broke open, the regime that it protected broke down.

These bifurcations are historical fact: they can no longer be averted. But this does not mean that the problem itself would recede into the past. The outcome of the bifurcations already underway needs to be managed: it is by no means decided. And there are other bifurcations on the way. The next wave will not be due to the sudden opening of semi-isolated or ideologically protected societies to the global flows of ideas, capital, technologies, and the similarly global markets of today, but to the unsustainability of the way modern industrial societies operate. In our own vital interest, the coming bifurcations must be anticipated and their unfolding consciously steered. This is a major challenge. As we have said: a working knowledge of bifurcation belongs to the essential knowledge of our age.

CHAPTER 2

Why Our World Is Unsustainable

The glittering coaches of the evening express speed along the tracks, pulled by their powerful engine. Inside the comfortable compartments people are talking, reading, playing cards, dozing; a mother is feeding her child; some young people are softly singing as one of them strums a guitar. Up front, the engineer glances at his watch. He is on time, and looks forward to the next station—and his warm bed.

Deep within the earth's crust, pressure is building up along a fault. Rock is grinding against rock, resisting for now, but not able to prevent the opening of tiny fissures here and there. If pressures reach a critical threshold, a tremor will shake the earth and its shockwaves will be felt for many miles. They will be strongest near the faultline. Not far from it is a gorge, and over it a high railway bridge. Its slender steel-and-concrete pillars are embedded in the rockwall and are solid—as long as the rock itself is solid.

Preliminary tremors shake small stones from the hillside near the track; the engineer glances at the rolling stones and looks back at the track. All seems in order there. Now his engine enters the last curve before the stretch leading on the bridge over the gorge. In the dining car the waiters clear away the dinner plates; sleepy passengers rub their eyes and begin to put their things together. The next station is not far down the line.

Below, pressure along the fault is building rapidly; the critical point cannot be far away. The ground under the long pillars trembles and shifts imperceptibly. The train enters the bridge that crosses the precipice. . . .

The scene is familiar; we have met with it countless times in books, movies, and television. We may be fascinated with the tension, yet we know that, unless it is a news report on an earthquake, real lives are not at stake, least of all our own. But is that still true today? Are we still mere bystanders or viewers, or are we more directly involved in the unfolding drama?

There are elements in this story that have a fearful resemblance to conditions in contemporary societies—and not only in the famed earthquake zones. Humanity is moving along a track of change in the belief (or at least the hope) of nearing reasonable goals of comfort and well-being. Those up front seem to be in control, even if we sometimes grumble about the way they run the engine. But deep underneath, various fissures have begun to develop. The surface signs are as yet limited, and it takes an unusually concerned engineer or passenger to notice the cascade of small stones shaken loose on the side of the track. Yet—we may be nearing the critical point where the structures that supported our lives and societies will give way. Will the engineer notice the signs of danger in time? Will he be able to stop the train and let us get off while we are still this side of the precipice?

Fortunately, real-world alternatives are not quite so stark. The choice is not between catastrophe and stopping, but between various kinds of evolving. There are many forks in the line ahead; we need not stay on the track we are on now. Our way can and must "bifurcate," yet the outcome need not be traumatic—we could shift toward alternative destinations, some worse, but others better than those we have thought to reach.

This assessment presupposes that we cannot go on as we have in the past. But is this assessment sound? Our glittering coaches are still on the track, the people up front claim to see nothing really unusual, no basic instability they could not control. This, however, should not give us a false sense of security. Like manmade trains running over hidden faultlines, nature also produces semblances of normalcy, sometimes up to the last moment before a catastrophe. Until almost the day it chokes life below the surface into extinction, the lily pond seems generously open to air and sunlight: if the lilies double every day, the day before they choke the pond they only cover half of it. Now, in the case of viable systems in nature, when a process reaches a critical point of stress or saturation, it veers off. Viable systems find a new way to exist and to evolve. But are systems in history viable systems of this kind? For it is they that face bifurcations, they that must find new ways to evolve.

We should not expect foresight and preventive action by complex but nonconscious systems, such as species and societies. For the most part these systems evolve by reaction: they mutate and transform when pressures have reached critical thresholds. We human beings, on the other hand, are capable of foresight. We could shift the evolution of our societies from the reactive to the proactive mode. We could perceive the forks in the road in time, and enter on tracks that do not lead to unsafe bridges over gaping precipices. Evidently, if we are to do so we need foresight. We must know if the stones that roll around us herald instabilities along the tracks on which we now find ourselves.

Why should we believe that our societies face major quakes and bifurcations? Why can we not go on in the future as we have been in the past? There are some basic statistics that can make this clear—that show that many of the proud practices of the Modern Age have now become unsustainable.

THE BASICS OF UNSUSTAINABILITY

A child born today in a country like the United States will require in his of her lifetime approximately 56 million gallons of water, 10,000 pounds of meat, 21,000 gallons of gasoline, 100,000 pounds of steel, and the wood of 1,000 trees. Consuming all this and more, he or she will produce an estimated 140,000 pounds of garbage. Children born in less "developed" economies will not require quite as many resources—and will produce somewhat less waste—but they will aspire to the "modern" lifestyle marked by consumerism, instant obsolescence, fast foods, fast-working medicines, and vertiginously changing fads and fashions—hence to a great many, and a great variety of, resources and pollutants.

It is not only that the children born today place—or intend to place—a great load on the global life support systems, it is also that there are more and more children. On the average, the world's population grows each day by almost 232,000 persons—each year by about 85 million. Of these newcomers, roughly 78 million are born in the poor developing countries and only 7 million in the still affluent areas. Due to this imbalance, already in 1983 three-quarters of the world's people lived in the Third world and only one-fourth in the rest. This proportion will shift to 79 vs. 21 percent in the year 2000, and 83 vs. 17 percent in 2020. If current trends continue, the population of Africa, the poorest of the poor continents, would triple in the 45 years between 1980 and 2025, growing from 500 million to 1.5 billion. But over 300 million Africans are already chronically malnourished, 150 million have acute food deficiency, and 60 million live on the edge of starvation—and nobody knows how many are dying of AIDS. Corresponding growth rates occur in poor countries on other continents. The population of Bangladesh is expected to double in the next 35 years, growing from 110 million to 220 million. How this could take place is not clear either: in that large but impoverished land every acre of cultivable land is already intensely cultivated, and millions live precariously on mudbanks, facing imminent starvation and disease. If allowed to hold sway, such growth trends would no doubt find their own termination in a sudden population collapse. Because the starvation syndrome is endemic and not likely to improve for decades, and because mortality from already contracted HIV-infection will be high for many years even if an effective preventive vaccine is found, the African population projection, the same as that for Bangladesh and other densely settled poor countries, is not realistic. It indicates only what would happen if the populations *could* survive. The number of those who *will* survive is far less than that.

Even at reduced growth rates, the human population will soon hover at the edge of the planet's carrying capacity. Carrying capacity, after all, is not what the Earth *could* carry but what it *does*, given the way contemporary economic and social systems operate. These systems are far from optimal. The world economy is critically inefficient: enormous capacities are left unexploited as the majority of the world's peoples are locked out of economic development. Many skills and great resources in creativity remain undeveloped. Instead of vibrant populations actively contributing to human well-being, uninformed and poverty-stricken masses become a burden on nature and a liability for society. In the short run, the situation is not likely to right itself. In the poor countries of the world there are always less and less funds for development. As the Third world attempts to service its more than $200 trillion foreign debt, it actually transfers an estimated $43 billion a year to the banks and governments of the "rich" countries and the international financial bodies created by them. The financial stringency imposed on the debtor countries squeezes every drop of liquidity from their economies, producing a spiral that leads to more debt, more stringency, and yet more debt. The number of critically empoverished economies is growing: in 1964 there were 24 countries on the United Nations' list of "least developed countries"; today there are 42.

Poor-country populations threaten to overflow the limits of sustainability both on the land and in the cities. In fact, city-living has become a major threat to the future. In 1950 only about 600 million people lived in cities; today there are well over 2 billion. In this regard, too, the imbalance between North and South is rapidly growing. In mid-century there were twice as many city-dwellers in the developed world than in the developing one, while today there are already more—one-and-a-half as many—city people in the South than there are in the North. Only one city in the South had a population of more than 4 million in the 1950s, yet by the turn of the century there will be 60. And by the year 2025 the number of poor-country megalopolises would rise to 135—if the world's urban carrying capacity could expand as much.

We are not doing much better in regard to producing the food required to feed the growing world population either. On the average, each second of each day we lose 1,000 tons of productive topsoil and 3,000 square meters of forest. As much as 35-40 percent of the Earth's land surface is threatened by desertification—this is an area equal in size to the United States, Canada and China combined. In China alone, the erosion of grasslands now amounts to 1.33 million hectares a year, with the total area of desertification affecting an estimated 86.6 million hectares—the third of all the utilizable grasslands. Despite great and much heralded efforts at reforestation, China's forests keep dwindling at an alarming rate. In areas such as Brazil, where peasants and young people cannot be massively mobilized for planting trees—and where great forests cover coveted pasture land—deforestation proceeds still more rapidly. The problem is worldwide. Over a third of the Earth's tree cover has already been lost. Our planet had an estimated 6.2 billion hectares of forests when serious human intervention began with the Neolithic Revolution, and it has no more than 4.2 billion hectares today. In our day trees disappear at a vastly accelerating rate; by the dawn of the next century we might be left almost entirely without tropical rain forests. This is already a cause of major environmental disequilibria. Trees not only provide much needed energy for cooking and heating, they also absorb carbon dioxide, the "greenhouse gas" of which we put 200 billion tons into the atmosphere each year. And they shelter myriad animal and plant species, of which we now lose more than 1,000 and possibly as much as 10,000, every year.

The report of a group of Chinese scientists, issued by China's Academy of Sciences in June 1989, identified the main elements of the problem. Entitled, "The Looming Threat to the Survival of China's Nation," the report concluded that "the rapid growth of population, the exhaustion of resources, the damage to the ecological system and the pollution of the environment are exerting a great pressure on the economy and will most likely become the biggest crisis for China's survival." The same could be said for all contemporary societies, indeed, for humanity as a whole. The planet's carrying capacity is reduced at the same times as it is becoming more populated. Rich and poor are becoming polarized and, despite superpower agreements on the limitation of nuclear weapons, on average five new nuclear devices enter the world's military arsenals each day. The world's governments spend roughly $1,000 billion annually on weapons and the military, and no more than a tiny fraction of that on health, education and social services.

The situation is not rosy. It is bound to lead to a reduced global carrying capacity of which the burden will be carried first of all by the poor. Growing deprivation by perhaps as much as three-quarters of the world population must ultimately lead to revolt and confrontation. The nation-states of our day are still fully equipped to make the worst of that.

THE ISSUES BEHIND THE STATISTICS

Statistics have to be regarded with caution: it is worth emphasizing that the load we place on the environment is not due simply to numbers. If humanity lived as it did 10,000 years ago, the Earth could support ten times the population we currently have, or are likely to have in the next century. What we do to the planet qualitatively is worse than what we do to it quantitatively. We overexploit nature's resources, and reduce its regenerative capacities. The effects of urban, agricultural and industrial pollutants are well known, but aside from outcries about oil spills and acid rain, their seriousness is seldom fully grasped. We now produce—and diffuse—an estimated 70,000 chemical compounds. Their impact on air, water and land is growing; in some cases it is already irreversible. The effects on the weather furnish a striking example.

If we feel that summers are getting hotter and winters milder, we are not mistaken. Our summers *are* getting warmer and longer. According to a study carried out in England at the University of Norwich, the years 1980, 1981, and 1983 were the hottest on record. Still hotter summers and milder winters will follow. Carbon dioxide (CO_2), produced by the burning of coal, oil, and wood in the atmosphere, now forms a blanket around the planet, reducing the amount of heat the surface can radiate into space. The greenhouse gases have accumulated to such an extent that world temperatures are already above the norm—0.7 degrees Celsius above it, to be exact. But this seemingly modest figure applies to the global climate as an average; near the tropics temperatures have risen considerably higher. If we emit CO_2 into the atmosphere at the projected (slowly slackening) rate, the Earth's average temperature would end up rising by a factor of 4 or 5 degrees Celsius.

A few degrees Celsius does not seem like much, but already a two degree rise would drive temperatures in the tropics five to ten degrees above the norm. The effects in the tropics as well as in more temperate zones would be far more negligible. World weather is a delicate and complex system; even small changes have large consequences. A mere 2.8°C cooling changed the prehistoric climate, depriving the dinosaurs of their essential wetlands; a 10°C cooling of the average global temperature would create a new ice age. Warming trends would have equally dramatic consequences. An increase of 1.5°C in the global average would recreate the climate of the first millennium A.D., when Nordic vikings settled a land ringed by green vegetation they named Greenland. At the 2.8°C warming level the North Sea would take on some of the characteristics of the Mediterranean, with balmy waters and palm-lined shores. While some cold spots would become vacation paradises, global agriculture would suffer. A 5°C warming would transform the U.S. Midwest into a dust bowl, and drop the water level in Colorado. As a result California would be deprived of water supplies as well as of hydroelectric power. Alaskan fish catches would increase thanks to warmer currents, while Lake Michigan would be evaporating and the wetlands of Louisiana would be lost to the sea. Elsewhere in the world the effects would be equally dramatic. The monsoon would miss the Indian subcontinent and irrigate the deserts of central Asia, and tropical Africa would dry out as water fell on the sands of the Sahara. The permafrost of Siberia would melt, but it would expose soils not able to sustain intensive cultivation.

There would be other just as serious effects on the environment. Even though greenhouse temperatures make for more of a rise in the tropics than at the poles, already a modest warming of the polar regions would melt some of the polar ice caps. The liberated waters would flow toward the equator, raising the level of the world's oceans. If the global average would rise by 1.5°C (2.7 degrees Fahrenheit), the sea level may rise some 20 cm. If there is a 4.5°C (8.1°F) warming, the waters would rise by as much as 140 cm (about 4.5 feet). The effect would be disastrous. More than two billion people—over a third of humanity—live within 60 km of a coast. While some of these densely inhabited regions are elevated, many others are near sea level. Low-lying cities and fields would soon be flooded. Unless dikes and levees were built in time, the skyscrapers of New York would project like islands in the sea, and London, Stockholm, Tokyo and dozens of capitals would have canals instead of streets. The changed weight distribution in the world's oceans would end by tipping the Earth, shifting its angle of rotation by a few degrees. This would be enough to change the location of the polar regions and wreak further havoc with weather patterns. Needless to say, not only agriculture, but all manner of social and economic activities would be severely affected.

The degradation of the environment would also create health hazards. The most widespread among them is likely to be skin cancer, caused by ultraviolet radiation. Normally buffered by the ozone layer, such radiation will be coming through in larger doses if the ozone shield is further thinned by chlorofluorocarbons in the atmosphere. The pumping of these gases—of which the ordinary aerosol can is a major source—could be reduced and even stopped, but the ozone layer could not be regenerated unless we had many more trees than we will have if deforestation continues.

But halting deforestation is not a simple matter. The loss of forests is directly linked both with the energy requirements of the growing world population—some 2 billion people still use wood as fuel for heating and cooking—and with the ongoing pollution of the environment. The burning of wood, the same as the burning of fossil fuels, leads to the accumulation of CO_2 in the atmosphere; this, as we have seen, interferes with the climate and changes the distribution of rainfall. Climate change in turn impacts on the growth and survival of trees. As new climates impact old vegetation zones, existing vegetation will inevitably suffer. While in the long term the extinction of one species of tree would allow new replacement species to immigrate, forests cycles tend to be long, and the immediate prospect is not bright. This is so especially in the tropics where trees have narrow tolerances: seasonal variations in temperature and rainfall are slight. Consequently tropical forests are especially sensitive to changes in water supply. Critical areas of the rain forests of Central Africa and Brazil could be decimated rapidly as a result of climate changes. Scientists at the Biosphere Project of the International Institute of Applied Systems Analysis (IIASA) found that this is what would happen if atmospheric CO_2 reaches double its present value. Not only the great tropical rainforests, even the boreal forests that ring the Northern Hemisphere, in a belt that is up to 1,900 km wide, could disappear: IIASA's projections show that almost 40 percent of the boreal forest belt would no longer be able to support the present species of trees. Given long (30, 50 year-plus) tree growth cycles, the warming trend could change conditions too rapidly to allow forests to regenerate by themselves—there would not be enough time for replacement species to migrate from the south, or for the current species to migrate north. Nor would reforestation be of help: seedlings of the current species that could survive at the time of planting would die before they would reach maturity, while the seedlings of the migrants could not survive under present planting conditions. In this light, almost the entire vegetation cover of the planet appears to be at risk.

It should be evident that it is not enough to prevent the chopping down of trees; also atmospheric pollution due to the burning of firewood (and of oil, coal and natural gas) must be cut back. Neither process is easily reversed. The majority of the world's economies still run on fossil fuels, as do almost all private and commercial vehicles. Accumulation of CO_2 can at best be slowed, not halted. Stopping the chopping down of trees runs into similar difficulties. Some 80% of the forests of the developing countries are used for fuel for cooking and heating. This process could only be halted if alternative energy sources became available—and affordable by the world's poor. Besides replacing fuel wood, the overexploitation of forests for commercial lumber as well as the clearing of forested areas for pastures would need to be stopped. Last but not least, the widespread practice of slash-and-burn agriculture would have to be eliminated: as much as half of the forests lost each year is likely to be due to this primitive method of food production. The forest cover is burned, and formerly balanced ecosystems transform into barren lands.

Reforesting the planet would call for an unprecedented worldwide effort. Technologies in agriculture, in industry and in the private sphere would have to change from polluting and wasteful to clean and sustainable modes; alternative energies would have to become available, and over 2 billion people would have to be brought from the clutches of extreme poverty into appropriately equipped economies. In the final analysis, the required effort presupposes a basic and radical shift in people's values and motivations. As long as suburbanites insist on riding private cars to work, shopping and recreation, the CO_2 content of the atmosphere will continue to rise; as long as people demand vegetables that have no blemishes, farmers will use pesticides on their crops; and as long as the average consumer wants to eat steak, and requires mahogany for his furniture, the world's forests will continue to be cut down.

The vicious cycle of deforestation and impacted weather patterns also reduces the global food produce. To feed the world's growing population great increases in arable lands as well as in yields per unit of land are needed. The authoritative International Rice Research Institute in the Philippines suggests that if a world population of 8 billion is to be adequately nourished, there must be a 50 percent increase in arable land in the next 50 years, together with a doubling of yield on the cultivated lands. The researchers are sanguine that this could be attained—trends and technologies being what they are. But the world's population may well grow to more than 8 billion in half a century; gains in land areas are likely to be greatly reduced through desertification; and climate change will probably produce unfavorable weather and hence smaller yields. The wave of optimism engendered by recent gains in food production leaves out of account that much of the gain is unsustainable: it is due to plowing highly erodible land that is incapable of sustaining cultivation over the long term. This is the case in China, in Latin America, as well as in sub-Saharan Africa. In the latter, the addition of 5 million head of cattle, sheep and goat is rapidly destroying vegetation and degrading the soil. There are serious shortages of water in India, and the same is on the horizon for China. The seas cannot replace what the land is losing: world fish catch has upper bounds of sustainability. When more fish are taken out than the seas can regenerate the size of the catch is bound to diminish. This has happened in many places already: for example, yields of the anchovy fisheries of Peru dropped from 13 million tons in 1970 to about 2 million tons at present.

Keeping pace with the still explosive growth of human numbers calls for restructuring the world's agriculture—for expanding arable lands and increasing the yields. This calls for reducing the pressure of human and animal populations, conserving topsoils, safeguarding the climate, and halting desertification. And these processes call in turn for reducing pollution and waste—hence for shifting from gross-growth oriented values and practices to qualitative and humane-growth-oriented policies and aspirations.

The energy problem constitutes another reason that the condition of contemporary societies is unsustainable. The energy crisis of 1972–1973 was dismissed as artifically induced, but the energy problem itself is not artificial. If it is not properly managed, an energy crisis may arise due to uneliminable natural causes. The belief that energy is a gift of nature and demand for it is a natural phenomenon is not tenable. Unconstrained energy demand will not be satisfiable in the future: historical rates of increase in energy consumption cannot be maintained. The reason is a combination of fossil fuel and fuel-wood depletion, and the hazards and costs of as yet insufficiently developed alternative energy technologies. If current trends would continue, the over 2 billion poor people who still use wood for fuel would deplete the world's fuel-wood resources even before the end of this century, while the more affluent populations of the modernized sector would use up the primary reserves of oil by the middle of the next. Lower oil prices produce the illusion that heavy crude reserves are, at least for the time being, practically limitless. But even if the total usable reserves would last until the middle of the twenty-first century, the economically exploitable primary deposits would be exhausted within a decade. And when secondary deposits are pressed into service, extraction costs will rise, and so will world oil prices. According to reports presented to the prestigious 1989 World Energy Conference in Montreal, price per barrel is expected to reach $30 in the year 2000—due to natural causes and not geopolitical ones.

A timely shift to renewable or abundant energy sources is imperative, and it calls for much further research and development and a flexible mix of the most appropriate of the new energy technologies. Nuclear energy alone, in its currently employed fission reactor form, could not solve the problem. Projections show that even if only the presently operating coal-fired power stations were to be replaced by nuclear power stations under the assumption of a moderate growth in energy demand, no less than one standard-size nuclear reactor would have to be pressed into service every 2.4 days, for a period of 38 years! Even aside from the staggering financial and technical resources this would call for, the nuclear option would entail significant risk factors. Standard varieties of nuclear fission reactors are hazardous and already outdated. Reactor safety is a problem even under peaceful conditions and with reliable and expert personnel. Under conditions of unrest, with potential terrorist acts and hostile actions, the problems are compounded. There is also the problem of disposing of nuclear waste, and of decommissioning aging reactors. Breeder reactors and reprocessing plants, while extending the energy output from fissionable materials, add to the risks: the plutonium cycle and the liquid sodium cooling system have further destructive potentials.

Research and development of the new fusion technologies could be important. Fusion may go a long way toward solving the energy problem, being comparatively safe, and making use of ordinary sea water rather than of hazardous and rare substances such as uranium or plutonium. The problems at present concern commercial application at a cost-competitive level. "Hot" fusion technologies are not far beyond the break-even point (where the amount of energy won in the process does not substantially exceed the energy that goes into running it), while cold fusion involves an as yet not well understood reaction that may or may not generate significant amounts of energy. Yet, even when properly developed, the nuclear fusion process could not serve as mankind's staple energy source without encountering a basic problem: that of thermal pollution. The laws of thermodynamics show that if one pumps large amounts of free energy into an open system, the system will move into a higher energy regime. This means that if significant amounts of energy are liberated from the nucleus of the atom and diffused on the Earth's surface, the atmosphere is bound to heat up and settle subsequently into a new thermal equilibrium. A higher global heat balance, whether produced by the greenhouse effect or by flows of nuclear energy, would trigger serious, and temporarily disastrous, global climate change. This problem does not affect technologies that make use of the energy contained in the solar radiation that reaches the Earth's atmosphere every day. Solar-based technologies do not add to the sum total of free energies in the biosphere but merely convert more of it to human use. Hence the need for appropriate energy mixes, even with the optimum development of the newest nuclear technologies.

Even safe and renewable energy sources have their drawbacks. Hydroelectric power dams, for example, have produced major ecological disasters, altering sensitive biological cycles both upstream and downstream and for many miles around. Technologies that directly convert sunlight into heat or electricity do not have immediate environmental problems, but they are forced to rely on the highly diffuse radiation that reaches our planet from the sun. It is difficult to see how they could power large urban-industrial centers, not to mention such emerging megacomplexes as the New York-Washington and the Tokyo-Yokohama area. To ensure large-scale output, areas the size of Arizona would have to be paved over with solar panels, or whole strings of geo-stationary satellites would have to be used to collect solar rays and beam them to receiving stations on the surface. And these large-scale energy-generating technologies are by no means free of the hazards of accidental or intentional misuse and abuse.

The factors that stress the contemporary world socioeconomic and ecological system are strongly interdependent. Here are some of the main causal strands and feedback loops:

- The burning of fossil fuels, deforestation, climate change, energy use, and the impact on agriculture are directly linked. Fossil-fuel based pollution creates acid rain, and acid rain kills trees. As trees die, the carbon stored in them is released into the environment. This adds to the CO_2 content of the atmosphere, reinforcing existing pollution and triggering climate change. Climate change in turn promotes forest die-back and depresses agricultural productivity.

- Urban concentrations, industrial complexes, misused or inappropriate technologies, and unsuitable forms of energy create pollution, cause deforestation, trigger climate change, and reduce world agriculture. Combined with the unrelenting growth in human numbers, they produce an unsustainable load on the environment. While the per capita load is higher the more affluent the population, the load imposed by extreme poverty also creates serious, and often irreversible, degradation. For example, people forced into subsistence modes denude vegetation in search of firewood and overexploit and erode marginal soils.
- Depressed environmental conditions aggravate the condition of the world's poor. As people are pushed toward the edge of subsistence, massive migrations are triggered: the number of refugees already exceeds 15 million. Mexicans are streaming across US borders, and transborder migration is expected to swell in coming years. Another wave could develop at the southern shores of the Mediterranean as starving North Africans take to boats to try their luck in Europe. A third major wave could advance from China into the thinly settled regions of Siberia. Many others could follow if the greenhouse effect leads to coastal flooding. Massive population movements of insufficiently prepared and resourceful people are likely to produce major stresses and a high level of social, economic and political tension.
- Environmental degradation and socioeconomic inequity have important long-term consequences. Poverty affects women and children most of all: in many societies when food is scarce, small children and women, not excluding expectant mothers, are the last to be fed. Babies born to undernourished parents are weaker and less viable than those born to healthy mothers. Furthermore, if these babies continue to be undernourished in the critical first year of life they may suffer physiological effects, including permanent brain damage. Starving children born of starving mothers will find themselves handicapped in later life when it comes to coping with the complex tasks and problems of contemporary existence.

There are no easy technological fixes to humanity's current predicament. Whether we take population growth, urbanization, deforestation, food, energy, or environmental pollution, the situation is the same: the track we are on leads to a precipice. Modern practices are not sustainable. Perhaps we ought to think of changing our values and aspirations, rather than of finding new ways to make nature satisfy them.

Humankind is not condemned to head passively to the edge of a precipice and into an abyss. There is a fork in the road, and with that the opportunity to change. There is no need to wait for crisis conditions before we do change; we should learn to read the signs in time and gather our wits. There are many alternatives put forward and tried out today, many new ways of living, of acting, of producing and consuming, of managing ourselves and our life-giving environment. We should weigh them and assess their consequences. It is up to us to change constructively, with humanism and freedom born of foresight and informed choice, rather than under heavy constraint, spawned by crisis and last-minute emergency.

CHAPTER 3

On Roots and Responsibilities

One day in the middle of March, 1944, the Hungarian children's humoristic weekly *Ludas Matyi* ("Mattie, the Goose-boy") came out two days early. As an eleven year old fan of the magazine, I was delighted to see it on the Budapest newsstands and bought my copy right away. As usual, I read its assortment of stories and humor eagerly, but was puzzled by the boldly printed headline which had no relevance to anything else in the issue. I remember showing it to my parents, and asking what it meant. They exchanged a worried look but did not answer. The headline, clearly visible at all the newsstands in the city, read "Hang On Matyi—Here Comes the Bend!"

At nightfall the same day, Hitler's army entered Hungary from the Austrian border and, by midnight, rolled into the capital. By 2 AM the small cream-colored cars of the Gestapo, the Nazi secret police, pulled up in front of the houses and apartments of influential Hungarians who were either Jews, or were known for their anti-Nazi leftist stance. Some time later, dozens of the Gestapo teams came back to their car, puzzled and empty-handed. Despite the assurances of dependable informers, many of the people they were looking for had vanished overnight. Those who had seen the headline of *Ludas Matyi* knew that the great "bend" was coming—and they were prepared to hang on.

There are various ways of hanging on in a bend. If one is on the priority list of the Gestapo, the advisable way is to vanish. If one is driving in a race, the best way is to brake before moving into it, and then press the accelerator to the floor. And if one finds oneself before the bend of a major bifurcation in society, the only way is to perceive the situation for what it is, get up one's courage, and prepare to make a real change.

The Nazi era arose in a critical instability in the Weimar Republic, and the bifurcation it created had almost succeeded in creating a reign of "Aryan supermen." If it were not for timely action on the part of the Allies, and countless acts of courage and foresight by those who were overrun by the Nazi machine but never gave up the fight, the Thousand-Year Reich would have been established and it may have lasted, if not for a thousand years, at least for several terror-filled decades, perhaps even to this day.

The first half of the 1940s was a crucial epoch; it posed a challenge to all who valued humanism and civilization. Fortunately, the challenge was perceived in the nick of time, and was met with determination and effective action. The enemy was clearly visible, and the means of fighting him were also evident. The 1990s are another crucial epoch, one that involves all people and all societies. But the enemy is not clearly visible, and what means we should use to fight our way to a successful conclusion is by no means evident. The matter deserves some thought.

ROOTS

Why do we find ourselves approaching a big bend in the road—a global bifurcation? Who or what caused the unsustainability of our condition? Where is the enemy?

The first culprit that comes to mind is technology. Modern technology has become a powerful force shaping our societies as well as our lives. Traditionally identified with the hardware produced in workshops and factories, technology today is seen as a complex system made up of people, organizations, role structures, skills and knowledge bases, as well as of material components. The technology-system is a combination of social systems and material artifacts dedicated to making things for consumption, as well as to making things for making other things destined ultimately for consumption—or for use in the service sector. Yet, complex and powerful as it is, technology has not become independent and autonomous. It is linked to the rest of society, and its "controls" lie ultimately outside of it. Its decisive milieu is made up both of the set of influential people and powerful institutions who decide what to research, to develop, and to produce, and of large numbers of ordinary people who choose from among the things offered to them in the marketplace.

It would be a gross oversimplification to lay the blame for the instabilities ahead of us at the doorstep of technology. Though frequently with unexpected twists and happy or unhappy side effects, technology did what decision-makers and consumers wanted it to do. People wanted faster transportation and more personal freedom of movement: technology delivered the motor car. They wanted more power to operate the cars and the countless gadgets that mean so much to them: technology delivered the electric power station and the required barrels of oil and tons of coal. They wanted a longer life, less risk of infant mortality—medical technology delivered as well. That the motor car produced urban pollution and traffic jams and poisoned the air of big cities; that oil- and coal-fired power stations polluted the atmosphere and nuclear fission reactors threatened the life of entire cities; and that a worldwide reduction in the death rate triggered a population explosion—these and similar consequences were neither demanded nor foreseen. To a young boy with a new hammer, Mark Twain said, everything looks like a nail. In our childish enthusiasm with ever newer, shinier hammers we have tapped left and right, high and low, and in the process built many things but have also broken quite a few. The problems we have created cannot be ascribed to a fault in our hammer; they were created by our assessment of what is a nail.

Opinion-molders and decision-makers did not grow much beyond the small-boy-with-a-hammer stage: they still profess deep faith in the powers of technology. Leaders in both government and business look at it as the key to national as well as corporate growth. In the late 1980s they have spent an annual $300 billion on research and development alone. At the center of their attention have been electronics, robotics and the information sciences, also nuclear technologies, aeronautics, advanced materials, and a wide array of genetic, chemical and bio-technologies. Weapons and weapons-related technologies have received the lion's share of the funding: an estimated $100 billion a year, fully one-third of the world total.

Indiscriminate tapping with ever improved hammers has been a major factor in making modern technology into what it is: a source of luxuries and creature comforts as well as of basic necessities—and of mind-boggling waste and pollution. New technologies based on high-powered research and development have destabilized social and economic systems and created unsustainable environments. But high-powered research and development, with few exceptions, drew its knowledge-base from science. Could it be, then, that science is ultimately responsible for our problems?

Science, in the classical view, is a search for knowledge pure and simple; it is neutral in regard to its consequences. This view has been seriously questioned lately, especially since the making of the atomic bomb and the advent of genetic engineering. We need only to recall the Oppenheimer case to perceive how complex the issue is—and how well taken is the attitude of those physicians, geneticists, and theoretical and experimental scientists who pose ethical questions in regard to all scientific activity, from basic research to the development and communication of the results.

Inasmuch as science fathered technology, and technology landed us in an untenable situation, science seems to be the culprit pushing us into a global bifurcation. But science did not create our untenable situation intentionally. Scientists are not evil geniuses out to destroy the world. They may have acted irresponsibly in claiming neutrality for their role in society, but they have not acted alone. Society took the knowledge they produced and turned it to its own purposes. Educational systems diffused scientific knowledge and interpreted it in light of the rationality prevalent in modern society; governments and businesses elaborated on it in light of their own needs and requirements. Everyone in society got into the act, not just scientists and educators, governments and enterprises, but also ordinary citizens, eager consumers of the spin-offs of scientific knowledge and technological know-how.

RESPONSIBILITIES

When we come right down to it, the real question is not the assignation of blame for wrongs already wrought, but the shouldering of responsibility for putting things right. Many actors and sectors have shaped our age, and many will shape the next. Science and technology, education, even art and religion, had a role in shaping our values and beliefs, and their impact will be more important than ever in the last decade of this century.

Ever since the inception of the Modern Age, the ensemble of the sciences have been major determinants of the way we think and act. We are not all scientists, and most of us do not know what particular scientific theories are all about, but the way we have been brought up, the way we look at the world, and what we see when we look, have all been subtly influenced by the rationality of modern science. That modern people are left-brain dominated and think linearly in terms of causes and effects, is due in large measure to the influence of a form of rationality that, though paradoxically already obsolete in the front lines of the contemporary sciences, has seeped deeply into the consciousness of our age. Its effects are many. They include a form of pragmatism that refutes to look beyond the surface—beyond what can be seen and touched, and bought, consumed, and ultimately discarded. Such pragmatism makes us unaware of distant and long-term effects, and irresponsible in regard to them. It leads to local efficiency and to global problems; to short-term benefits but long-term crises.

Though a powerful agent of the evolution of the Modern Age, modern science is not the only force that made this age into what it is. Strange as it may seem on first sight, art has been, and is, an equally powerful agent. We are not all artists any more than we are scientists, yet art is subtly influencing how we perceive, what we feel, and how we relate to each other. Art, after all, is not limited to museums, galleries and concert halls; it is all around us, in the shape of the houses we live and work in, the form of the products we use, the tunes we hum, the novels we read, and the tragedies and comedies we view on television and on the movie screen. Our sense of beauty, and our everyday desires and ideals are constantly shaped by the perceptions that come to the fore in both the "pure" and the "applied" arts. That we did not become unfeeling robots, mindless computers molded by our conception of scientific rationality, is due in no small measure to the constant presence of art in daily life.

Religion is a third and no less important force shaping the mind-set of an age. It would be wrong to think of religion either as a set of superstitions that our scientific mentality should finally overcome, or as the sole guiding light of our time. Religion is neither a superseded nor a dominant component of the Modern Age; instead, it is a vital and integral one, in company with science and art. Our sense of ultimate meaning and significance, our conception of what is truly important and valuable, and even that sense of the sacred that was so strong in all pre-modern societies and that we have not entirely lost even today—these are all molded and given form by the belief systems of the world's great religions. We may not adhere to any doctrine or visit any church, synagogue, or temple, yet we share Christian, Judaic, Moslem, Hindu, Buddhist, Taoist, Confucian or some other religious, mystical, or mythical values and world views.

Last, but of course not least, our age has been shaped by the institutions and methods of education. Education is not in itself a source of perceptions, values, knowledge, and modes of behavior; it is merely a transmitter of them. Yet even in that capacity, education has had a major impact on thinking and acting in our day and age. The reason is that educational systems, no matter how broad, can be only limited channels of transmission. They cannot transmit all the values and beliefs of an age, and what they select for transmission gains singular emphasis. That we take things apart if we want to know them, that we overvalue specialization, that we profess no responsibility in regard to future generations, that we view ourselves as different from, and somehow better than, any other nation, and that we think we are separate from, and a cut above, nature—these are all consequences of the way we have been taught in schools, and the way our personalities are shaped by informal and continuing education in later life.

Science and art, religion and education had a crucial role in shaping the Modern Age, and they will have a crucial role in shaping the next age as well. If they are to live up to their epochal responsibilities, science and religion, as well as art and education must become conscious of their role and aware of their societal impact. In this regard they have a long way to grow.

Take science. Even if there have been highly visible outcroppings of social consciousness among scientists since the first atomic bombs exploded over Hiroshima and Nagasaki, the attention of the representative majority of the contemporary science community has remained steadily centered on fields that are either well funded, or have narrow esoteric interest. Except for areas that are well endowed with research grants, such as those connected with cancer and AIDS research and with what is euphemistically called "national defense," scientists follow only their own highly specialized noses and distance themselves from the concerns of their society. In universities and academies, the teaching and researching of science are compartmentalized into specialized disciplines without regard to their relevance to anything beyond the disciplinary bounds. The majority of scientists has grown apart from human concerns: even from workers in other fields. Scientists have become super-specialists, encased in ivory towers built on the protected turf of prestigious institutes and universities.

The situation is not any better in regard to art. The artists of our day have become even more aloof than the scientists. In many artistic circles, *l'art pour l'art* has become a sacred tenet, not to be compromised on pain of excommunication. Artists create their works in the rarified atmosphere of studios where no newspaper, no television program, no herald of the vulgar affairs of the everyday world is allowed to penetrate. Art historians assess painting, sculpture, poetry, drama, music, dance, and the other fields and branches of the arts as though they evolved exclusively under laws of their own; laws laid down by artists of genius and modified only by other artists of equal genius. Theoreticians of art analyze art as a relation between art "object" and individual "perceiver" and critics are so preoccupied with technicalities and style that they seldom stoop to considerations of social impact and relevance. High art has divorced itself from popular art and, though with some notable exceptions, left the matter of social relevance to the latter.

Organized religion is nearly as introverted as the specialized branches of contemporary science and the individual domains of contemporary art. Churches, temples and synagogues are more concerned with their own integrity and with their power vis-à-vis other religious groups than with the soul of their brethren. Religious strife is a recurrent phenomenon in history, and parochialism still clouds the efforts of the major religions. As the Gulf crisis has demonstrated, "holy wars" are still being fought between "believers" and "heathens." Parochial rivalries fire animosities and cause conflict and violence in many parts of the globe.

Education is not guilty of losing sight of social relevance, but it *is* guilty of falling seriously behind in conveying the knowledge and ideas produced by the leading scientists, artists, and humanists of our day. Present-day educational establishments still behave as if the world could be neatly divided into independent and sovereign nation-states—into "my country" and the rest of the world. Our educational establishments also divide the systems of knowledge into the categories of nineteenth century science, projecting a world picture fragmented into physical reality, the living world, and the sphere of human intent and action. The resulting division, into a scientific-technological, a human- and social-scientific, and a humanistic-spiritual culture, makes an integrated outlook with a healthy holistic perspective all but impossible.

The epoch-making role of science, art, religion and education can no longer be responsibly exercised by the unreflective impact of introverted practitioners. Science and technology, the same as art, education and religion, did not deliberately create the nonsustainable situation of our day and need not be blamed for our predicament. But, even if not accepting blame, they must yet accept responsibility. The communities of technologists, scientists, artists, religious denominations and educators must divest themselves of their narrow interests and join forces in the building of a new, more humane and sustainable world.

And what about today's leaders in government and in business? Are they not at fault as well—and do they not share in the responsibility for leading the parade down the path to the precipice?

Aurelio Peccei, the Italian industrialist and founder of the Club of Rome, has put his finger on the problem when he said that today, precisely when mankind is at the peak of its powers, it lacks the wisdom to put its powers to proper use. The problem is equally acute in the private and in the public sectors. Politicians and corporate executives do not know how to make proper use of their corporate and state powers. Of course, using state power to increase national status and wealth, and using corporate power to increase growth and profitability in business, is generally considered a proper use. But in an epoch of bifurcation this is not necessarily so. If the system behind the state and the corporation is not sustainable, propping it up merely postpones its downfall and makes it that much more violent.

National or corporate collapse via global trauma is not the intention of today's leaders. But their current efforts may yet produce just these results. A case in point is recent events in China.

In spite of the hard line taken by the Deng regime after the June 1989 student demonstrations, optimism among the populace remained surprisingly high. Although there was ideological *Gleichschaltung*, there was also more food on the market, and the peasants—the great majority of the Chinese population—were especially pleased: not only did they have enough to eat, they were also getting richer. Neither the peasants, nor the workers, nor yet the bureaucrats worried about where the leadership's current policies would lead. Yet, where they were leading China was precisely in the opposite direction—toward ruin and not toward prosperity. The long-term outcome will inexorably be collapse rather than stability. It does not require a complex systems model to see why.

In mainland China, 22 percent of the world's population lives off 7 percent of the world's cultivated land. Clearly, the land is intensively worked. The peasants have their own parcels and can sell their produce on the open market. They make a little money and, like good businessmen, they reinvest most of it in their enterprise. They put up greenhouses—simple plastic structures hung on bamboo or iron frames—so that now they can grow cabbages and other vegetables for the greater part of the year. They have more produce, and thus more money in hand. But they also need more hands around the farm, and the one-family-one-child system cannot provide for this. Female babies tended to disappear, as couples tried their luck a second and a third time. Now the regime permits couples to have more children if they are willing to pay additional taxes. The peasants have the money and they need the children, so they make babies and pay willingly. Soon more people will till the land, covering more areas with intensive cultivation. They will squeeze more produce out of marginal soils.

The result is not difficult to anticipate. Soil erosion, as noted in Chapter 2, is a serious problem in China; it impacts fully one-third of that vast country's utilizable grassland. At the same time environmental pollution is spreading from cities to vast rural areas. Some 40 percent of China's rural enterprises have become the sources of heavy pollution in a previously bucolic countryside. And there is a growing shortage of water that forces limits on irrigation and on household and sanitary use.

What will happen when more people till the lands, feeding more mouths and producing more babies, is entirely predictable. The likely scenario goes like this. For a while, the growth of the rural population-curve rises sharply upward. The rate of basic food production also curves upwards, but then soil erosion, pollution, and insufficient irrigation takes its toll. The curve inflects. If current (basically just minimal) levels of food intake are to be maintained, China's population growth-curve should inflect at the same time. But, unlike cabbages on the marketplace, the people looking for cabbages do not just disappear from one day to the next. They grumble and go hungry; then they revolt. The process is familiar to China in her 5,000 year history. Autocratic rule has always been interspersed with the revolt of the suppressed masses. China's current regime replicates the policy of the Emperors; and the same as the foregoing dynasties, it digs its own grave. This could be a deep and tragic grave indeed, for China's has over 1.1 billion people.

The scenario of Chinese development is but one of countless examples of public- and private-sector shortsightedness in maximizing short-term benefits while risking long-term collapse. Whether in the countryside or in the cities, whether in agriculture or in industry, our socioeconomic systems are operating close to stability thresholds right now. These thresholds can be rapidly overstepped, and then the systems may suddenly collapse.

The problem is that nearly no one, whether in the public or in the private sector, is willing to face long-term costs for fear of losing short-term benefits. Today's leaders trade long-term sustainability for short-term advantage. What does it help, politicians admit, to achieve benefits beyond one's term in office when one's successor may reverse such responsible policies? The planning cycle in government is from election to election; and in business from one meeting of the shareholders to the next. If pressed, businessmen tend to repeat Lord Keynes' dictum: in the long term we shall all be dead. It may be well to remember, however, that change is accelerating and time is shrinking; we may be facing the "long-term" within our lifetime. And if not in ours, then certainly in that of our children. No one is glibly offering that our children—and our nations and corporations—should by then all be dead.

It is time for businessmen, no less than politicians, to factor the human condition into their day-to-day decisions. They know that the bipolar world of military and economic power is giving way to a multipower world of many competing actors; that overall economic growth is slowing; that the quantity and quality of information is increasing; that the pace of technological innovation is accelerating; and that surprises and uncertainties of all kinds are mounting. They also suspect that these are indications that the period of extensive, quantitative growth of the postwar years is giving way to intensive, qualitative development. What they must realize as well is that qualitative development, unlike quantitative growth, produces fundamental change in the system in which it occurs. Extensive growth can occur by accretion—more of the same can be accumulated—but intensive development is not cumulative; it is transformative. Sooner or later it is bound to bifurcate the system that manifests it.

The unsustainability, and hence the coming bifurcation, of our socioeconomic systems has deep roots. The leadership of contemporary societies cannot undo or reverse the historical trend, but it could make the bifurcation to which it leads less sudden and traumatic. Doing so would constitute an exemplary objective. As Plato said, all people seek immortality through their deeds and accomplishments. Politicians and business executives, no less than scientists, artists, religious leaders and educators, want to be remembered for having contributed to the progress of humankind. After all, such contribution is perhaps the only valid ground for a sense of immortality, and the only basis for feeling fulfilled in one's professional life. If today's leaders were to factor the wider context of the human situation into their day-to-day decisions, they could look back on their career and say, "I have done my share to prepare human affairs for a new and positive future." In the absence of such concerns they can have no assurance of having done anything more than delay the coming bifurcation, and render more difficult its day of reckoning.

PART TWO

AN EVOLUTIONARY OPTION

CHAPTER 4

The Nonequilibrium Crystal Ball

The modern world is unsustainable and on the way toward a bifurcation. A new world can and must be created. But what new world? The age that follows the modern age is hardly ever identified. People just call it the "new" or the "post-modern." It seems clear that the next age will not be the same as the one we live in, and that it will come after it. But what will it be like in its own right? That question no one seems able or willing to answer.

The desire to identify the post-modern age comes up against a major problem: that of seeing into the future. Gone are the days when people could content themselves with consulting sages, astrologers and soothsayers; tea leaves, horoscopes and crystal balls tend to become fuzzy when it comes to answering queries about the future of humanity. Social scientists, too, seem reluctant to commit themselves. The reason is that in the standard branches of the social sciences one can read and extrapolate trends provided the parameters are constant, that is the epoch itself is stable, but not if the rules of the game themselves change. Periods of fundamental change bedevil the calculations.

To see the future in a period of bifurcation calls for an evolutionary science. It is already taking shape. Though classical crystal balls remain cloudy, the "nonequilibrium" crystal ball begins to glow. It is the tool of the new science of complex systems—systems that evolve both in nature and in the human sphere in conditions far from equilibrium. In these nonequilibrium conditions, systems are dynamic: they balance their unstable structure through many self-regulating and self-organizing processes. Being unstable, they are frequently unpredictable. Thus the nonequilibrium crystal ball does not foretell a ready-made future. It tells only of what one can predict—which is important, even if it is not everything.

The nonequilibrium crystal ball is worth gazing into; let us begin our session.

To begin with, we ask two simple questions. *Can we predict the human future? And if so, within what limits?* The limits of predictability in the human sphere may not be the same as those that apply to simpler systems. Take a well-wound clock, for example. Its hands move across the dial face with strict regularity. If we know where the hands are now, we shall know where they will be five minutes, one hour, or twenty-four hours from now. Also the movement of the planets in the solar system is, for all intents and purposes, regular and dependable, and hence predictable. But the "movement" of humanity through history may not be as predictable as that. One can even doubt that it is predictable at all.

The human future is predictable if the past is, that is, if there are laws or factors that have determined the course of history. Could there be such laws, and if there could, what would they be like?

Two kinds could possibly enter into play. One pertains to the nature of the human organism, the other to the nature of societies. The former set of laws or factors are biological, and if they are determinant they would create a kind of biological (more precisely, genetic) determinism. The latter set are sociological, and they in turn would spell social (that is, socio-cultural) determinism. Let us look at each in turn.

BIOLOGICAL EVOLUTION

If biological factors determined the course of history, they will also determine the future. Our history, the same as our destiny, would be decided by the biological evolution of our species. Our future will be unchanged if our species remains unchanged, and different if our species evolves. This viewpoint accords with a reputable school of thought that looks on the information encoded in our genes as the ultimate determinant of behavior. Sociobiology, as developed in the 1970s by Harvard biologist E. O. Wilson, produced an impressive array of evidence. The central principle is that individuals tend to behave in a manner that maximizes their inclusive fitness. "Fitness" is measured by reproduction, by the success of individuals in projecting copies of themselves—more precisely, of their genes—into succeeding generations. Genes, in the view of biologist Richard Dawkins, are "selfish": their sole purpose in life is to re-create themselves. The complexities of the human body and of human behavior are only the means that serve the success of this endeavor.

If we press the argument to its logical conclusion, even social interactions appear largely determined by our genes. Other than the embellishment of this or that function through the creation of this or that social structure, human society is as much the expression of the genetic endowment of its members as an animal or insect society is. We may think that we live in societies freely created by us; in reality we live in super-anthills and ultra-beehives in which the complexities of structure and function are due mainly to our own genetic makeup. Our genes make us egotistic: social structure is the result of a trade-off between the selfish aims of individuals and the recognition that many of these aims can be better served by joining forces than by going it alone. Our genes make us aggressive: the history of societies is the history of wars, only interspersed by the cessation of hostilities because periodically there is a need to recoup one's strength and regroup one's forces. Our genes make us thirst for power: the structures of society are the product of the power struggle of individuals as the stronger subdues and binds the weaker. And so on, in regard to the basic traits of human behavior: they are all mirrored by corresponding characteristics in society.

The conclusion one draws from this view is that human society is unlikely to change in the near future. People will be selfish, aggressive, power-seeking, and all that they are in addition, tomorrow just as they are today. There will be wars, power-structures and the rest in the future as well. Human society, the same as the human body, is the way it is because human genes are the way they are. As long as our genes remain the same, society remains the same. No hope for a different future, at least not in the span of the next generations. It would take a new human to make a new society, and the new human awaits a new mutation in the evolution of our species.

There have been many dreams of creating a new and superior human being; they ranged from Friedrich Nietzsche's *Übermensch* and Hitler's Teutonic superman to more recent speculations concerning the control of human breeding stocks through eugenics. The Nazi regime attempted to diffuse the Aryan stock and eliminate what it considered inferior breeds, such as Gypsies, Jews, and Slavs. Extermination camps that put the worst excesses of medieval inquisition to shame were the means of accomplishing this "final solution". More well-meaning champions of genetic engineering now speak of the elimination of "deficient" traits through splicing the amino acid sequences that make up the genetic code of individuals. They hope that laboratories could soon produce a species genetically superior to today's sapiens, with traits such as greater intellectual ability, less proneness to aggression, fear, and rage, less susceptibility to disease, and tolerance of a wider range of climates and environments.

The prospect appears promising. Through careful genetic manipulation, controlled interbreeding, and the selective diffusion of the new stock we could mutate sapiens to a higher form. We could breed *Homo supersapiens* whose selfish traits would be balanced by genes coding for sociability, whose aggression would be kept in check by an instinct for belongingness, and whose hunger for power would be mitigated by a genetic disposition for cooperation. *Supersapiens* may not need a larger cranium and bigger brain than today's sapiens: it would be enough that he uses more of his brain than sapiens does. Intelligent, sociable, and cooperative, *supersapiens* would create a new society and bring in a new age.

New genes, a new man, and new society—all made to order. The concept may be attractive; the more is the pity that it is hopelessly unrealistic. Why so? *First*, because we have only the vaguest notion of the specific DNA structures that would produce particular personality traits. Our knowledge falls short by far of the requirement to mass-produce the new traits seriatim, like a brand of dishwasher. *Second*, because creating individual personality traits would not be enough in any case: we would also have to "legitimize" the emerging traits, assure that individuals possessing them can reproduce and spread them. In the absence of a radical intervention in the normal processes of society, the "new man" with his hyper-intellectual and nonaggressive disposition would soon end up in the dustbin, eliminated in the competition with more egotistic and aggressive specimens. The latter would breed on, making more of their selfish and aggressive kind. *Third*, because the personality traits of individuals do not uniquely determine the nature of the order that comes about when the individuals interact. The social system does not merely mirror the traits of its members: unselfish individuals will not necessarily make unselfish societies any more than nonaggressive individuals will make peaceful societies. Sociable persons could yet be poor organizers and managers; peacefully disposed individuals could create high-stress situations with which they are poorly equipped to deal. A social whole is never the simple, linear sum of its parts—one cannot reduce the characteristics of society to the sum of the characteristics of its members. *Fourth*, because of the time it takes for genetic change to unfold. A period of 50,000 years is fast in biological evolution; it is an evolutionary "leap." Yet this period is fully one half of the entire span of existence of sapiens, from its emergence in Africa to our day. If we assume that we can accelerate the process by conscious intervention, we would still have to count on at least 30 generations passing before a mutant gene could diffuse and define the dominant traits of our population. This would give us a quasi-instantaneous evolutionary leap of some 6,000 years. While 6,000 years is the merest wink of an eye in biological time, it is far too long to be of relevance to the future of humanity. When confronted with these realities, the dream of creating the future by creating the new man soon evaporates.

There is no need to grieve over this: it does not matter in the least that we cannot mutate our species as we would like. The kind of change we need in the future is not the kind that our good sapiens stock would be unable to produce. We have been essentially the same genetic individuals for the past 100,000 years and, except for straightening our posture, reducing the size of our jaw and increasing the size of our brain, and developing a better gripping hand (and less well-gripping feet), we have not been very different for the past 5 million years. Genetically, we are surprisingly close to the higher apes and almost identical with a whole series of previous hominid types—no specimen of which would be enjoy having as our next door neighbor. We— "*homo sapiens sapiens*"—have produced a whole series of cultural types within our own history. Only 5,000 years have elapsed since the advent of *Homo classicus*; 1,000 years since *Homo medievalis*; and 400 years since *Homo modernus*. Each of these cultural types created a different age, even if their genes were the same. For it is not genes that determine the nature of an age. The genetic heritage of sapiens is generous enough to give rise to many scores of ages and societies, several times more than those that have come about in our history. *Homo modernus* could be followed by many a *Homo post-modernus*.

One day, probably in the as yet distant future, our species may very well mutate biologically. But we should keep this event distant rather than try to bring it near: it is filled with danger. A mutation—*any* mutation—has such a high probability of depressing the viability of a species as to amount almost to a certainty. Only a long process of natural selection can weed out the unfit mutants and find that small fraction that has enhanced viability. In human society such selection can no longer be natural: any observed change in the gene pool is likely to be manipulated by genetics and medical science and selectively treated by people and institutions. Presently, no conceivable development of genetics can assure that favorable mutations will be artificially produced. All interference with the human gene pool—even those that seek the elimination of genetic "defects"—can entail grave dangers and are correctly viewed with suspicion.

Widespread mutations could not only be produced intentionally: they may also be involuntarily triggered. The prospect of an accidental mutation is real—as real as that of any other technological catastrophe. With the higher radiations levels to which our organism is exposed, with the vast intake of pollutants as we breathe the air, of chemicals as we eat our food, and of synthetics even in our clothing, the isolation of a gene from the rest of the body is seriously under attack. Experiments show that genetic changes can and do result from radiation and unnatural living conditions. If they could result from the conditions to which modern men and women are exposed throughout their lifetime, the result would be almost certainly negative. While we do not know how to create and spread a desirable mutation, we are fully capable of creating an undesirable one—given the likelihood that any mutant we create will turn out to be undesirable.

If accidental mutations occurred in large numbers, the gene pool would soon be severely contaminated. Future generations would be born with defective genes; they would have lesser resistance to disease, shorter lifespans, and fewer and just as defective children. For all intents and purposes, the effects would be irreversible: just as we do not know how to intentionally create a mutant that has more desirable characteristics than those we have today, we do not know how to intentionally create a mutant that regenerates our present characteristics. The conclusion is obvious: we should let biological evolution be. We should not try for purposive mutations, and avert the specter of accidental ones. But we need not grieve over our inability to create a genetically new human. What we need is not a biologically, but a culturally mutant sapiens.

SOCIAL EVOLUTION

What about social evolution, then: is it predictable or is it not? Few people would agree that society could be as determined as the hands of a clock. There are, however, other kinds and degrees of determinism, and just how, and to what extent, society may be determined is the subject of lively debate. On the one side are the philosophers and social theorists who believe that society is governed by "iron laws"—laws of history that will determine its future as they have determined its past. On the other extreme are those thinkers and scientists who deny any kind and degree of determinism in regard to society. Far from moving ahead on a predetermined trajectory like the hands of a clock, society has no trajectory but makes its way through chance and circumstance.

Let us look at the determinist hypothesis first. The future of society is predictable if there are determinant factors of sociocultural evolution and we know what they are. The factors could be iron laws, natural principles, or even the will of God. We could know them through the empirical method of science, through mystical intuition, or through religious revelation. All that counts is that the determinant factors should exist and be knowable. If we know them, we can predict the future.

Determinism of this kind lands us in a fatalistic frame of mind. The future will be what it will be; as a once popular song had it, "que será, será." We may want to know what the next year or the next century will bring, but this interest will stem more from curiosity than from a desire to master our destiny. Predicting the future will be like solving a crossword puzzle: the solution exists already, the task is just to find it.

Yet predictability of a complete, fatalistic kind is hardly ever affirmed in the sciences and only seldom in the world's religions. Almost always some leeway is perceived for conscious and purposive action—for intervention even in an otherwise deterministic process. Even Marxist doctrine, the radically deterministic theory of historical materialism, allowed purposive human action to influence the course of events. And non-Marxist doctrines are far less deterministic than that. Many scientists hold that human action can not only move along, or temporarily block, the realization of a given type of society but may be entirely decisive of the choice of the kind of society that will come about. History, the positivist social scientists say, knows no deterministic laws. It is "just one darn thing after another." Society, like Topsy, was not "made" but just "grew." Understandably, then, history is full of surprises. For example, Tsarist Russia yielded to the Bolshevik ideology even though Russia was not a bourgeois society and had no proletariat to speak of—not to mention a historically conscious one. The intellectually sophisticated Germany of the Weimar Republic gave rise to Hitler, even though Nazi slogans and theories bordered on the insane. Iran's Shah, the self-styled King of Kings with a powerful military and police machinery at his command, fled before the followers of an exiled and ancient Islamic fundamentalist. And the same kind of thing repeated in Battista's Cuba, Marcos' Philippines and the wild swings of allegiance in Ethiopia and Benin . . . to mention only a few of the major "surprises" of our century.

Historians did not predict and politicians did not anticipate these and similar events. It is with good reason, say the positivists, that our age is known as the Age of Surprises.

Yet the fact that unexpected events occur at times does not mean that we must renounce the idea that history follows laws of its own. The surprises could, after all, be due to our *ignorance* of the laws. Moreover, there could be laws that do not strictly determine what will happen, but give only probabilities and indicate overall trends. Such nondeterministic (so-called "stochastic") laws are known in natural science—physics would be lost without them—and they are likely to hold true in the human sphere as well. There could be patterns in history, even if there is no full determinism in the way it unfolds. The patterns would apply to large ensembles of events; to the overall envelope within which individual events find their unique and seemingly haphazard appearance.

PATTERNS IN HISTORY

What kind of patterns would there be, then, in the welter of events in history? The possibilities are not as vast as we might think. The following make up the basic set; anything further is likely to be an elaboration of one or another among them:

- *the circular* (monotonously cyclic) *pattern*;
- *the helical* (innovatively cyclic) *pattern*;
- *the linear* (directly progressive or regressive) *pattern*; and
- *the nonlinear* (statistically progressive or regressive) *pattern.*

The Circular Pattern

The basic variety of the circular pattern suggests the mythical concept of "eternal recurrence." The future is not entirely new; essentially it is a repetition of the past. This was the dominant conception of change in early pastoral and agrarian societies, where it was inspired by the seemingly eternal recurrence of the seasons. In Western intellectual history the concept was revived in the nineteenth century by the philosopher Friedrich Nietzsche, but has found few adherents in recent times.

Those who would maintain the validity of the circular pattern often look to the history of China for support. There, this pattern seems to have held true for thousands of years. From the inception of the first Chinese dynasty in 221 B.C., until the revolution that brought down the last in 1911 A.D., Chinese society did not seem to change significantly; it repeated the selfsame pattern over and over again. Periods of social and political integration under a powerful dynasty were followed by periods of disintegration under the impact of external invasion or internal revolt. Disintegration yielded in turn a new integration as another dynasty came to power and gathered the scattered segments into a new integral unity.

The factor that permits one even to entertain a circular approach to history is the stability of the environment—of the social, political, technological, economic, climatic, ecological and human environment. In the few instances where such a view of history has been applied, the total environment could be seen to be relatively unchanging, virtually static. But if such stability ever existed in places like ancient China, it certainly does not exist in the modern era. The very richness of human activity and interaction mitigates against historical repetition. While Schopenauer's famous warning that people who forget history are doomed to repeat it may have some cautionary validity, our experience is that different people in different circumstances reacting to different stimuli behave in different ways. As events in Russia and Eastern Europe can testify, even politicians are continually surprised by the morning headlines.

The Helical Pattern

The cyclical pattern has been further elaborated in recent times and is no longer thought of as a strict recurrence of past events. Even if the general sequence of events recurs, some historians hold, it does so in a new form. This concept underlies the "great cycles of history" theories. They have illustrious exponents, the most notable among them the Renaissance jurist and philosopher Giovanni Vico, the nineteenth century historian Oswald Spengler, and his twentieth century counterpart Arnold Toynbee.

Vico, in his 1725 opus magnum *The New Science*, set forth a conception according to which all cultures follow a basic cycle termed the *corso*. Cultures develop in response to the needs and desires that correspond to specific times within their cycles. Even if they borrow ideas, institutions and values from other cultures, nations, and societies, they borrow only those that correspond to their cycle-specific needs. The major stages in the cycles Vico called the *heroic*, the *religious*, and the *philosophical* (or scientific). The third, highest state is always followed by a period of decline and decadence, leading to the initiation of a new cycle in the framework of a different culture. Each cycle ends with individuals seeking mainly their own interests, engaging in the pursuit of pleasure without living up to their civic responsibility. Thus, having run their *corso*, cultures disintegrate—unless they learn to make use of the New Science to liberate themselves.

The thesis of the disintegration of societies at the end of their natural cycle was revived in Oswald Spengler's famous study *The Decline of the West* (1918-1922). Strongly influenced by the ideas of Nietzsche, Spengler suggested that cultures, like individuals, have their own life cycles. They go through stages of birth, growth, maturity, and final senescence. His ambition was to write a "morphology of history"—a comparative study of cultures. The cultures Spengler described were those of Egypt, India, Babylon, China, classical antiquity, Islam, the West, and Mexico. Each of these "powerful cultures," he said, leaves its imprint on mankind as it goes through its own life cycle. In the final stages of the cycle it produces a "civilization"; a conclusion that follows from a previous process of growth. Having produced its civilization, each culture enters on the path of decline, as the West was supposed to have done in Spengler's own time.

Spengler's ideas have in turn influenced another key feature in historiography: Arnold Toynbee. As he recounts, he read Spengler's *Decline* in 1920, and the concept of a plurality of civilizations with each following its own cycle made a deep impression on him. Toynbee found significant parallelisms between the history of ancient Greece and Rome and that of modern Europe: World War I appeared to him as a reprise of the Peloponnesian War or the Punic Wars. In *A Study of History* (1934-1954) he expanded the parallelisms into the concept of a universal civilizational cycle called "the tragic pattern." He applied the grand scheme of the tragic pattern to some thirty civilizations; he described each civilization in terms of thirteen concepts showing how they pass from growth to breakdown.

The difficulty with this view of history is that it is predicated on an arbitrary division of historical epochs. Eras rise and develop, and then end on specific dates in specific places via specific events. This conception has become as outdated as the view of the biological cell as a pristine, isolated entity that transports food stuff in and waste material out. Living organisms are so integrated in their environment that the boundaries between the organism and its neighbor, between it and its environment, are understood to be murky and variable—an arbitrary artifice of the research laboratory. And in much the same way the boundaries between people and between historical epochs cannot be determined by pronouncements, treaties or lines drawn on someone's map. The fundamental premise of the cyclical view of history—that on the day following a battle, a treaty, or a declaration, all the people within earshot are different and completely oriented to the new regime—has been shown to be a myth. Indeed, in the widespread debates of the 1950s, Toynbee's cyclical view was generally discredited, much as Spengler's concept of cultural life cycles has been 30 years earlier. Today, few historians would subscribe to a cyclical interpretation of history, although more and more of them search for the recurring patterns that underlie the manifest course of historical events.

The Linear Pattern

The linear-progress concept of history is a dominant concept of modern times. To conceive a definite directionality underlying a seemingly chaotic welter of events requires either a knowledge of the more distant past, when things were significantly different from what they are at present, or a rate of change that is sufficiently rapid to be readily perceived. Today we have both, while traditional societies had neither the one nor the other. They could seldom conceive that conditions would change in any unalterable and irreversible fashion. Even during the Middle Ages linear progress in history appeared improbable; if there was such progress it could only be tied to individual destiny as it headed toward salvation rather than damnation. But when modern science emancipated itself from the Judeo-Christian doctrine and gave birth to modern technology, the concept of progress penetrated public consciousness. Technological advances attendant on the first industrial revolution inspired great euphoria, and the concept of historical progress as linear, that is continuous, smooth and assured, seemed firmly established.

The modern concept was predicted by the Marquis de Condorcet already in 1795. In his "sketch of the intellectual progress of mankind" the French aristocrat announced that all the causes which contribute to the improvement of the human species must remain forever active and their extent must forever increase. Civilization has always moved, and always will move, in a desirable direction. When Darwin's *Origin of Species* appeared in mid-nineteenth century, technological optimism was further reinforced. The kind of views preached by the Marquis de Condorcet were "scientifically" proven: progress was enthroned as desirable, true, eternal, and inevitable. Technology would improve the conditions of life year after year, and with the improvement of the quality of life there was likely to be an improvement in the quality of those living as well.

The technologically inspired linear progress concept has been thoroughly ingrained in modern society. It was badly shaken, however, in post-war years. The development of the atomic bomb, the occurrence of technological catastrophes like Three Mile Island and Chernobyl, and major negative impacts on the environment such as acid rain, urban pollution, oil spills and the thinning of the ozone layer, have conspired to weaken its hold. In the last few years there has been a growing tendency to conceive of history in the contrary mold of linear *regression.* Young people and intellectuals now profess some form of technology-inspired apocalyptic pessimism: we shall deplete the environment, overcrowd our cities, fail to halt the arms race, and sooner or later a major catastrophe will befall all of us.

The concept of enduring and basically linear progress—or regress—occurs also in the spiritual domain. Regress is represented in all doctrines that view man as guilty of the original sin and hence as fallen, expulsed from paradise. By contrast the Jesuit biologist-theologian Pierre Teilhard de Chardin combined Christian doctrine with biological evolutionism in a linearly *optimistic* view: evolution is bound to lead to a higher and higher stage of spiritual and even physical evolution. Humanity's evolution is due to a process of "convergence" or "totalization": a kind of compression through the formation of increasingly tight relations among increasing numbers of people and organisms on a finite planet. Ultimately we shall form a Noosphere around our planetary matrix—a single, organic unity, enclosed upon itself and co-extensive with Earth itself.

On the linear view of historical development we can have an extrapolation of history into the future that leads to a heavenly Utopia; and an equally valid extrapolation that leads to a fiery, hellish end. Which future is true? In point of fact: both and neither. Toward which end the world is headed (the very concept is specious—what happens the next day?) will not be known even the day before. The day before the world ends in catastrophe could easily be the most idyllic day the world has ever seen; the day before universal peace and joy could just as easily be our darkest. There does not seem to be much sense in viewing the path to the future as a linear process.

The Nonlinear Pattern

The nonlinear conception of historical development is a very recent development on the intellectual scene. While notions of historical development as oriented in some direction and yet interspersed with forward leaps and sudden regression are found in Western as well as in Eastern myths and philosophies, the recognition of nonlinear change as a basic feature of evolution in both nature and history had to await the emergence of the evolutionary sciences of nonequilibrium systems in the 1970s and 1980s.

The basic idea was foreseen by the philosopher Alfred North Whitehead more than half a century ago. "It is the business of the future to be dangerous" Whitehead wrote in his 1933 *Adventures of Ideas*. "The major advances in civilization are processes that all but wreck the societies in which they occur." Whitehead could not explain the crisis-bound nature of progress: all he could say was that, although we have an intuitive premonition called Historical Foresight, we do not know enough of scientific laws to be able to predict the future even a year hence. But our knowledge of scientific laws has made great progress in the last 60 years, and the leaps and bounds of historical development are now explained by the new sciences of nonequilibrium systems.

In the framework of the new theories, human societies, the same as biological species and ecologies, are particular varieties of complex systems arising in the constant flow of energy in the biosphere. They evolve through multiple bifurcations. These intersperse long periods of stability, and crown the peaks and valleys and seemingly random oscillations of the epochs of instability. Underlying these processes there is a general directionality, a long-term trend which unfolds from earliest prehistory to our own day, and onwards into the future.

The unfolding of the evolutionary process is strongly nonlinear. There are numerous fluctuations and reversals, and many periods of stagnation. Major perturbations, such as wars and social, political and technological revolutions, rock and ultimately destabilize societies. Governments fall, systems of law and order are overthrown, new movements and ideas surface and gain momentum. There is a period of chaos as new orders take shape. But new orders do arise, and history sets forth its jagged course from the Stone Age to the Modern Age—and then beyond.

SEEING THE FUTURE

Whenever patterns are perceived in a process, there is the possibility of extrapolation. Whatever the nature of the pattern, it provides a handle for grasping something about the way it will unfold in the future. This is true of each of the here reviewed patterns: they can all be extended beyond the Modern Age to say something meaningful about what will come after it.

Of course, for the circular pattern the future holds nothing fundamentally new; it is a repetition of the past. But according to the innovatively repeating cycle there is some novelty in recurrence; each cycle moves society along a given axis, vague as the fate of future cycles may be.

The implication of the linear pattern is more decisive: it is either outrightly optimistic or pessimistic. If history progresses forward the future will be a utopia; if it regresses our fate will amount to a dystopia.

Extrapolation from the nonlinear pattern is not as simple. The new sciences suggest that complex nonequilibrium systems evolve in a definite direction, even if they do so in sudden spurts and with frequent surprises. The overall trend is nevertheless toward societies of increasing size and complexity, of increasingly high and numerous levels of organization, of greater dynamism, and of closer interaction with the environment. This means that, according to the "nonequilibrium crystal ball," post-modern society will be globally integrated and technologically advanced. Human settlements will be organized on multiple levels, from the grass roots of villages, farming communities, and urban neighborhoods, through townships, districts, provinces, national and federated states, all the way to the global community as a whole. Each level will be coordinated with all the others. And the globally integrated network of human societies will also be integrated with the globally integrated system of the biosphere.

This vision is not a prediction. The nonequilibrium crystal ball does not foretell what *will*, only what is *likely*, to happen. The laws of social evolution are not deterministic; they remain open to surprises. The same as the past has been, the future may be beset by reversals and deviations. Some of them could be serious. If the next reversals include a thermonuclear war, or a major irreversible degradation of the environment, there will not be any future at all—the planet will become uninhabitable. Yet even such an ultimate catastrophe would not contradict the known processes of evolution: biological evolution, too, always leads to the disappearance of some of the already evolved systems. In fact, almost 99 percent of all the biological species that have ever emerged on this planet have now become extinct; and a large proportion of the culturally specific human groups and societies that arose in history have also vanished. Only the extent and the time-scale of a future catastrophe would be new. Rather than involving one type of system, such as an organic species, an ecology, or a human socio-cultural group, it would involve all of humanity and all of the biosphere, and it could last not for centuries but for untold millennia.

Drawing on the findings of the new sciences of complexity, we can now identify the post-modern age with a little more specificity. Ours will be a global society, integrated yet diversified, dynamic and complex, and organized on many levels, from the grass roots to the global. But, we must add, it may or may not come about in reality.

The last proviso sounds unsatisfactory. On a moment's reflection, however, the nonlinear extrapolation still turns out to be happier than the principal alternatives. Unless we are comforted by the notion of a preordained destiny, we will be pleased that this scenario harbors more freedom for human action than a deterministic unfolding of history. Unless we are adventuresome to the point of foolhardiness, we shall be content that the scenario is more determinant than the fully random sequence of historical events of the positivists. And unless we are afraid of novelty, we will find this scenario more interesting than a mere cyclic recurrence of past phenomena.

That there are no guarantees that the global society of the post-modern age will actually come about can only give us incentive to gather our wits and act. It is, after all, up to us to learn to go with the trend and to *bring* it about.

CHAPTER 5

The Third Strategy

THE BUTTERFLY EFFECT

The global age of humankind, as other ages before it, will be born in the creative womb of chaos. Like all evolutionary transformations in complex systems, it will be a product of what has come to be known as the "butterfly effect."

Just what is the butterfly effect? It was originally discovered in the 1960s by U.S. meteorologist Edward Lorenz as he was modeling the world's weather on one of the largest computers then available. The weather system of the globe, he found, is in a permanently chaotic state. This means that it is impossible to predict which way it will evolve: its trajectory is sensitive to the slightest alterations. A small change here or there, and the evolution of world weather shifts unpredictably from one of the outstretched "butterfly wings" of the so-called chaotic attractor to the other. (See Figs. 1 and 2.)

It is not without reason that long-term weather forecasts often turn out to be wrong. It is almost impossible to predict the evolution of a system in a condition of chaos. The longer one extrapolates into the future, the greater the uncertainties. Making the best of a difficult situation, meteorologists and chaos theorists came up with a picturesque interpretation (based on an ancient Oriental notion) of the butterfly effect. It is, they say, the effect of a monarch butterfly in California that unexpectedly flaps its wings: the atmospheric turbulence this creates produces an entire series of bifurcations—and next week's weather in Outer Mongolia becomes completely unpredictable.

The fact is that in a condition of chaos, the slightest modification can expand and change the dynamics of the whole system. This is not necessarily a negative factor: there are many instances of chaos, and some of them are highly creative. The brain, for example, has neural networks that are in a constantly chaotic condition. Because of this they can respond to the finest, most

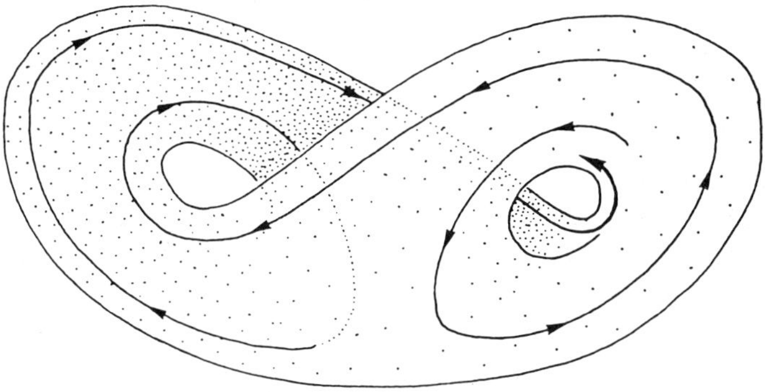

FIG. 1. The original model of air currents in the atmosphere by Lorenz. While the attractors are determined, the future of a trajectory within them is unpredictable by an observer. In fact, the trajectory is so erratic that Lorenz despaired of predicting the weather by simulating it with this dynamical model. Unpredictability is a generic characteristic of trajectories defined by chaotic attractors. (Reprinted, by permission, from Ralph H. Abraham and Christopher D. Shaw, *Dynamics; The Geometry of Behavior*. The Visual Mathematics Library, Santa Cruz: Aerial Press, 1984–85.)

minute changes in their input. Cognitive states that are unusually close to chaos can be especially creative: scientists and artists, poets and prophets often conceive their finest ideas and receive their greatest inspirations in the seemingly unordered "altered states" typical of meditation, dreams, and trance—or during particularly stressful periods in their lives.

Chaos can harbor danger as well as promise. As we have seen, world weather, being in a state of chaos, is sensitive to minute variations. Some of these variations can be artificially produced, and they can have unexpectedly harmful effects. Propellants escaping from spray cans, for example, though seemingly insignificant, can add to the blanket of greenhouse gases that surrounds the planet, and this small increment in temperature can produce significant changes in the global weather system.

Society, too, enters a chaotic state from time to time. This is not a state of anarchy but of ultrasensitivity—the prelude to change. In a chaotic condition, society is sensitive to every small fluctuation, to every new idea, new movement, new way of thinking and acting. In the final analysis chaos in society spells human freedom—freedom to change the structures and institutions in which people live their lives. The heavy hand of the past is lifted, and

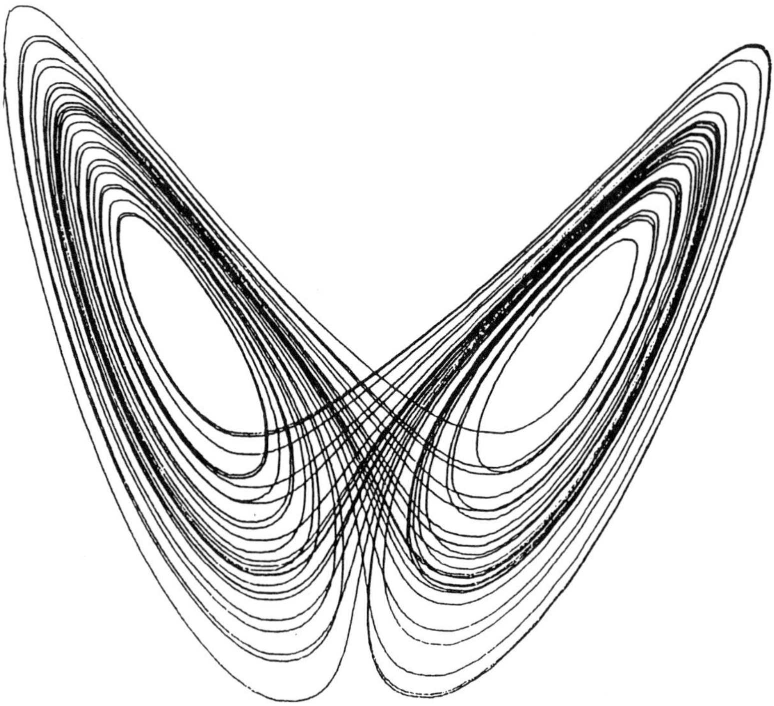

FIG. 2. A more recent computer drawing of the Lorenz attractor, with its elegant butterfly shape in evidence. (Reprinted, by permission, from Ralph H. Abraham and Christopher D. Shaw, *Dynamics: The Geometry of Behavior*. The Visual Mathematics Library, Santa Cruz: Aerial Press, 1984–85.)

individual creativity has room to unfold. Not dictators, armies and police forces, but the changing values and ideals of people are the butterflies that, flapping their wings, determine which way society will grow and develop.

We can and must make productive use of the butterfly effect as we enter the age of bifurcation. The human condition is uniquely suited to this end. People are relatively stable and conscious "parts" in unstable and nonconscious "wholes." Unlike society itself, the individual members of society are capable of thinking, planning, and envisioning alternative courses of action. This is a unique condition. In natural systems the parts are not conscious, and they cannot influence the destiny of the wholes they form.

THE SPECIOUS ALTERNATIVES

If we are to make proper use of the unique condition we humans occupy within the social, economic, cultural and ecological systems of this globe, we must know not only *that* we can decisively influence the course of system change, but also know *how* we should influence it. Finding strategies to achieve change with lasting benefits poses a major challenge for our times. The classical objectives of progress and development have become obsolete and even dysfunctional.

Take, for example, the time-honored objectives of liberals. They have always been intent to remove constraints on the freedom of individuals on the hopeful premise that when an individual acts so as to maximize his own interests, he also maximizes the interests of his society. What is good for one, is good for all. Individuals can pursue their interests however they perceive them: an "invisible hand" harmonizes even selfish motivations with the public good. This is the foundation-stone of the policy of liberal laissez-faire, and it is no longer a solid one. In times of stability, when the various strata of society could develop in a fairly equitable fashion, free competition and the market mechanism can distribute benefits without need of major interventions. But in a period of instability and rapid transformation, this may not be the case. And if it is not, if the invisible hand atrophies, this is a problem. If there is no policy available to cope with the inequities, instead of a caring hand, we shall find a tactless foot that gives painful kicks when we least expect it.

The classical objectives of communism became just as obsolete, and far more evidently so. In communist societies the automatic coincidence of private and public good did not dominate; a single party, equipped with the proper "historical consciousness" has led the way, designing the institutions of society and prescribing the roles and tasks of individuals. But in practice the ideologically inspired objectives of the party seldom corresponded to the majority's ideas of what is good and desirable. Furthermore, the party-created structures tended to be inefficient and corrupt. No wonder that people, locked into frustrating niches in giant state machineries, eventually rebelled.

The "First world" of liberal democracy believed in the invisible hand and scaled down the public sector to the bare minimum. The "Second world" of communism feared the invisible foot and built up the public sector to an omnipresent maximum. The "Third world" of less developed countries vacillated uneasily between these options, swinging with few exceptions toward the laissez-faire one. But we now know that neither strategy is functional. The fallacy of the Marxist system became evident in the dramatic events of 1989 and 1991, as one communist party after another was challenged and fell apart. The fallacy of the liberal system is becoming evident as well, although not quite as dramatically. In laissez-faire societies the private sector had acquired a high concentration of wealth, and dominant economic, social, and even political influence. The result has not been freedom and autonomy for individuals in a condition of socioeconomic well-being, but a highly competitive environment where the winners live in mansions and the losers on the street, and where both rich and poor are threatened by the alienation of city life and the waste and pollution of irresponsible affluence.

The assumption that everybody must be either liberal or communist—either on the political "right" or on the "left"—is false. Both these strategies have been tried, and regardless of what benefits they may have brought in their time, their hour has now passed. Human societies need to find a more timely concept.

Individual freedom and autonomy, the same as social and economic justice and equity, are perennial values of human life. But the classical strategies whereby freedom and equity are sought in practice have become dangerously outdated. Societies have transformed fundamentally since the political programs of liberalism and communism were formulated. The program of liberal laissez-faire was valid in seventeenth century Europe when new technologies created radically changed conditions that made the rule of absolute monarchs obsolete and unacceptable. Under those circumstances the injunction "he who governs least governs best" made eminent sense—it created a much needed space for individual freedom and initiative. But in the late twentieth century freedom is no longer threatened by the power of hereditary rulers. A further adherence to the program of laissez-faire may produce out-of-control conditions that threaten the well-being, and perhaps the very survival, of many populations.

The Marxist strategy of communism was intended to rectify the shortcomings of the laissez-faire system. It was to ensure economic and social justice for the empoverished peasants who, in the heat of the first industrial revolution, were forced off the land and into the factories and sweatshops of the new "captains of industry." But in the late twentieth century, the way to achieve social and economic justice is not by nationalizing the property of industrial overlords and feudal landlords and placing power into the hands of a single political party. This only leads to bureaucratization, inefficiency, and corruption. No wonder that, when glasnost opened a crack in the power of the party system, the winds of change, instead of rejuvenating the regimes, blew them apart. In the ultimate decade of this century communism as a state doctrine passes ineluctably into history. If a government would persist in the communist strategy of centralization, the society would be likely to break down under the combined weight of inefficiency, inertia, and corruption. But eliminating the specter of communist centralization is not the panacea it is often said to be: if government would persist in the classical strategy of liberal laissez-faire, society would still suffer, not because of inefficient centralization but because of the noxious side effects of uncontrolled and uncoordinated individual initiatives.

If liberalism and communism were our only options, the situation would be truly desperate. We would not even have the classical dilemma of whether it is better to be dead than red, or better red than dead: we would be dead on either option. Fortunately, these are not true alternatives. The solution is neither to be red nor to be dead, but to be alive and evolved.

THE HUMANISTIC EVOLUTIONARY STRATEGY

The classical strategies of liberalism and communism, being no longer functional, must be replaced by a more timely and functional "third strategy." The new strategy is to optimize individual freedom and autonomy at the same time as ensuring social justice and equity. It is to co-evolve the person and the society. We cannot halt the evolutionary process or regress it to some prior stage. We must "go with the flow" but can choose which way we go. The bifurcations that await contemporary societies allow diverse outcomes. There is no law of nature or of history that would preempt the decision as to which of the many forks along the way society will take.

There are several evolutionary forks, in addition to the ever-present possibility of devolution into chaos and anarchy. There are, after all, many ways to create a dynamic, technologically advanced, integrated and diversified multilevel society. Such a society could be a *hierarchy* commanded from the top and forcing its many parts and elements into a pre-designed unity; or it could be a *holarchy* where the diverse parts and elements participate in setting goals and objectives and cooperate in carrying them out. Humanity possesses the technologies—the organizational skills and the hard- and soft-wares of interpersonal communication and consultation—to create an evolutionary society based on voluntary cooperation born of understanding and solidarity. But it also commands the technologies to create a global society that locks individuals into predesigned roles and niches, and surveys and controls their actions, even their values and motivations.

The laws of evolution, in nature as well as in history, are probabilistic and not deterministic. They are permissive. While they do not permit everything (if they did, they would no longer be "laws," but mere chance), they leave a great deal of latitude in the outcomes they do permit. The basic alternatives are chaos and anarchy along the devolutionary fork of a bifurcation, or a global, multilevel, dynamic, diversified and integrated system along the evolutionary fork. Within the evolutionary fork the choice is between social evolution at the expense of individual development, or the co-evolution of the individual and society. It is up to people to decide whether their society is to be devolutionary or evolutionary, and if the latter, whether it is to be a hierarchy or a holarchy.

In the ultimate decade of this century—which is likely to be also the ultimate decade of the Modern Age—we have the option of the co-evolution of people and societies. We could choose the optimum path toward an evolutionary future, avoiding the fallacies of Marxist communism as well as those of laissez-faire liberalism. We could exercise the necessary restraints to channel the processes of economic, social and political globalization into humanly manageable pathways; and at the same time we could create the interconnections needed to ensure the kind of coordination without which our globally extended world could no longer be kept on a safe and sustainable course.

The co-evolution of the individual and society is a tall order, but not a utopian one. It is high time to give the humanistic variant of the evolutionary strategy serious thought. We still have a window in time; a precious and perhaps nonrecurring chance to think and to prepare. The ideas and visions we now produce could be the butterflies of the ultimate decade. It is up to each of us to flap our wings—and to launch our bifurcating societies along a humanistic evolutionary path.

CHAPTER 6

A Vision for 2020

> Consider the following task. You are to design a "third strategy" that goes beyond the classical and now obsolete political doctrines of liberal laissez-faire and communist centralism to create a humanistic evolutionary scenario. The target date is the year 2020. You can assume that bifurcations in the 1990s have created an opening for fundamental change; by the first decades of the twenty-first century new ideas have a realistic chance of translating into social reality. What are the essential elements of the new strategy? What are its principal goals and objectives? To start the flow of ideas, here is one candidate for a humanistic evolutionary scenario—one "vision for 2020."

The basic objective of the third strategy is to launch humanity on the path toward a global holarchy where human beings can co-evolve with human societies. This calls for maintaining mastery over the complex and interdependent world we have now created. The globally extended interconnections that have evolved in the Modern Age are, and will remain, necessary components of the post-modern world. But they must serve rather than dominate humanity. They should become the instruments for effectively managing ourselves in harmony with each other, and in harmony with all other systems of life on this planet.

In view of this basic consideration, the humanistic and evolutionary "third strategy" has two sets of objectives, distinct but interrelated. The first set is essentially *defensive*: it is to avert the evolution of the structures of society at the expense of the individual. The second set is *pro-active*: it is to build up, and make effective use of, the connections that link people all over the world with each other, with their environment, and with the biosphere as a whole. The former is to *safeguard the development of the individual*: this requires that we restrain and control the evolution of hierarchically oriented political and economic systems and processes. The latter is to *create a global-level holarchy*: a network of cooperative relations in fields and areas where worldwide coordination is useful, and indeed imperative.

THE INDIVIDUAL SAFEGUARDS

Objective Number One: Restrain the Power of Nation-States

The development of individuals cannot and need not be planned: it need only be permitted. The first requirement of a humanistic evolutionary strategy is that it create space for personal growth and creativity. This means a strategy of restraints in areas where the evolution of hierarchic structures and institutions poses a threat to the freedom and autonomy of the individual. One of these areas is political by nature but is in fact more than political in everyday reality. It is the myth of the modern nation-state, with all its entailments including its controls, its structures, and its claims of sovereignty.

In the contemporary world national sovereignty has become almost sacred. As in the USSR and in Yugoslavia, the governments of national states must be pushed to the very brink of chaos before they would relinquish sovereign powers, either to subnational entities such as cantons, provinces, regions, republics, and states, or to transnational bodies such as regional federations, economic communities, or international and global organizations. But, no matter how natural it may seem, this unyielding adherence to national sovereignty is inscribed neither in the laws of society nor in the laws of human nature. It is an historical product, and it must pass into history when the age that produced it has passed.

In its legal and institutional form the modern nation-state dates only from the Peace of Westphalia, concluded in 1648. The concept became institutionalized throughout Europe in the seventeenth and eighteenth centuries and has spread to the far corners of the world in the great wave of decolonization after World War II. While developing countries objected to almost every concept they inherited from their former colonial masters, they never contested the validity of the principle of sovereign national states. As a result, today's world community consists of some 180 nation-states, and merely a handful of territories with non-sovereign status. Humanity has accepted the inter-*national* system as a permanent feature of the world.

This has to change. Following the global bifurcations of the late twentieth and early twenty-first centuries, there will not be compelling reasons to maintain the world system of national states; by the year 2020 people in many parts of the world will have a chance to create new types of social, political, and economic units. At the grass roots level they could form human-size communities where the voice of each person could be heard. Other systems could manage the economy, cultivate the cultural heritage, protect nature—and protect one society from aggression by another. There will no longer be a need to place all powers of decision into the hands of central governments—as Croatians and Estonians well know.

There are no constraints in the psychology of individuals that would limit their allegiances to monolithic national states. No one needs to swear exclusive allegiance to one flag, and to one flag only, and wave it in the conviction that it stands for "my country, right or wrong." People can be loyal to several spheres and units of society without being disloyal to any. They can be loyal to their community without giving up loyalty to their province, state, or region. They can be loyal to their region and also feel at one with an entire culture, and with the human family as a whole. As Europeans are Englishmen and Germans, Belgians and Italians as well as Europeans, and as Americans are New Englanders and Texans, Southerners and Pacific Northwesterners as well as Americans, so people in all parts of the world have both narrow and broad identities—even if the latter are underutilized and atrophied because of the myth of the nation-state.

By 2020, neither the grass roots nor the broader cultural and human identities of people would need to be neglected in favor of such a unilateral system of allegiance. Contemporary Europe is not a nation-state and it already provides a wider identity for Europeans. The Europeanness of Europeans is neither a source of confusion nor a ground for conflicting allegiances. Indeed, if the English and the Germans, the Belgians and the Italians would not persist in the legal fiction of forming sovereign nation-states, their Europeanness would unfold that much more. New England, Dixie, and the Pacific Coast are not nation-states, but people identify themselves with these units in addition to identifying themselves with the United States as a whole. If the federal government would not persist in claiming nation-state sovereignty for itself, the regional identities of its people could evolve more, and come better to the fore. Americans would not be any the less good citizens for that.

In the next thirty years, humanity could grow to be both global and local. It would have the chance to develop the international system in the form of a holarchy, on many decentralized yet coordinated levels at the same time. The highest level would be the global, and the lowest would be far more human-size than most of today's nation-states—smaller, more participatory, and more democratic.

But first, the question we must face is, what are we to take as "human-size"? The optimum size for human habitation has been debated for centuries, but few if any of the historical concepts have remained applicable. For example, Plato's ideal of a city-state of about 500 people is far too small in the highly populated world of today and tomorrow. His main concern, that the limit that a man's voice can carry should be the limit of his community, is now superseded: modern communications techniques can carry the voice of individuals over great distances and enable large and even dispersed groups of people to interact with one another. More recent speculations about optimum size have also tended to be unduly restrictive: Ebenezer Howard's city of 32,000 would still be too small in a global age. In a world equipped with modern communications technologies, and having upwards of 10 billion people, many more people can and need to share urban environments. But even the best communications technologies cannot overcome the problems of urban megacomplexes—crime, overcrowding, high cost of living, impersonal living and working environments, and a tight and competitive job market.

Studies of "livability" carried out in North America and in Europe show that few cities above a population level of 500,000 can provide optimum living conditions. Around that size, however, the urban environment combines economic, social and cultural advantages with the benefits of manageable distances and a sense of community. It would be wise if, wherever possible, the people of the early twenty-first century would refuse to aggregate into cities much larger than a million. The rapidly growing megalopolises of the late twentieth century would be best scaled down by attracting people to the country and to satellite communities of manageable size. Provided that the myth of the nation-state is surrendered, there will be no need for large capitals, whether for fueling national pride or for supporting the infrastructures of national governments—and, incidentally, assuring the comfort and convenience of the power elites.

Cities, of course, are not the basic social unit. Even if almost half the human population will be urban by the end of the century, the growth of cities is an historically recent phenomenon. It need not be prolonged into the post-bifurcation epoch. Vast populations have already come to the realization that they do not want to live in urban environments; with suitable economic and settlement policies they could get a chance to live and work in the country. Even where cities would persist they could be surrounded by farms, villages, and small and medium-sized towns in social systems that have diversity as well as unity. A truly human-size community embraces urban as well as rural environments and links the diversity of both settings for the benefit of its people.

In the next century the optimum size for urban-rural communities is likely to be somewhere between the dimensions of the classical city-state and the modern national-state. A community that exceeds 60–80 million is likely to be too large. The periphery could become detached from the capital, diversity could interfere with unity, and structural disequilibria could appear between rich and poor, city-dwellers and country-folk. Social systems of more modest dimensions would have better chances of providing a humane environment; they are more likely to maintain themselves with a degree of organic unity. Historically, people who survived with such unity were well within the 60–80 million population range, for example, the English, the French, the Dutch, the Finns, the Austrians and the Hungarians, to mention only a few in Europe. Also national subcultures fluctuate around this dimension—New Englanders, Texans, those of the Pacific Northwest and of the Maritime Provinces, to draw on some examples from North America. Even the Chinese and the Indians, enormous as their national populations are today, have evolved as regional cultures of relatively modest size and have consolidated into nation-states and grown to giant dimensions only in this century.

The Soviet Union and Yugoslavia today; China and India tomorrow . . . the decentralization of super-large nation-states through the downward transfer of sovereignty is a major goal of the third strategy. But it is not the only goal; if the strategy is to safeguard the freedom of individuals, related goals must be pursued in the sphere of practical politics.

Objective Number Two: Restrain the Power of Politicians

Politics—the regulation of interactions in an organized community—is a perennial requirement of humanity. But politics, as well as politicians, can serve many ends. They can serve the hereditary sovereign, the militant dictator, or the elected leader. They can serve powerful lobbies and special interests. Even if well-intentioned, they can be misguided by incorrect perceptions and incomplete information. But politics can also serve the genuine interests of the people, even if historically this feat was not frequently accomplished.

Political systems often start with noble ideals, to end as self-serving authoritarian systems. Goverance tends to degenerate into a fight for political power and personal advantage. National political machineries have a vexing tendency to turn into tsardoms of one kind or another. For example, the United States, though a bulwark of democracy, assigns awesome powers to the President; Watergate did little to diminish them. The President is the supreme commander of the armed forces, can appoint his own cabinet and can determine the country's domestic as well as foreign policy. While a system of checks and balances keeps him from abusing his power, and national public opinion exercises further restraint, his day-to-day power remains great and at times imperial.

The United States, of course, embodies moderation compared with most other countries, especially in the South. In Mexico the president is more of a tsar than the head of a democratic state. During his six-year term of office enough power is concentrated in his hands for him to be not only the undisputed political leader of his country but also one of its richest citizens. The situation is worse in many Central American and African countries. The main concern of the top leaders is often to gather as much wealth and influence as is necessary to stay in power and, when that is no longer possible, to exile themselves to a life of luxury.

Restraining the power of politicians may sound utopian: power, as everyone knows, corrupts and corrupt politicians have no intention of voluntarily surrendering even a fraction of their power. Yet limiting political power is not unfeasible: it does not call for changing the nature of politicians, only the nature of societies. And fundamental change in the social sphere is occurring already, and more changes will come by the year 2020.

Suppose, then, that by this date states will be created that do not claim absolute sovereignty for themselves. If democratically structured, such communities will have leaders with lesser pretensions of power and fewer illusions of granduer. Democratically elected leaders would return to their civic or professional occupations after a limited term in office. And during their tenure, they would make use of modern communication systems to consult the people of their communities on major issues, rather than taking decisions on their own.

Direct democracy is feasible in small or medium-sized states where people are closely in touch. In such communities there can be small and flexible administrative systems with delimited tours of duty. Serving in their political posts could be seen as a civic duty, much like serving on a jury is seen today. There could be safeguards against accessing illicit powers and illicit gains. Community goverance could be effectively protected, if not from all forms of corruption, at least from the more virulent forms of power abuse.

THE GLOBAL HOLARCHY OBJECTIVES

Restraints alone, though needed to safeguard individual freedom and development, will not be enough to achieve a humanistic evolutionary future. Even if the world's social and political systems could be successfully decentralized, connections among the decentralized units would soon grow and intensify again. Given our history, and our technologies of production, trade, marketing, transport and communication, the global level in human affairs can no longer remain underdeveloped. It would be futile to try to halt the globalizing process—we could no more undo global flows and processes than we could uncook a half-cooked egg. But it would be just as foolish to precipitate the process of globalization and force bifurcations on unprepared societies. This is what decolonization has done in the post-war years, and glasnost forty years later. The humanistic evolutionary strategy is neither to indiscriminately open societies to the embrace of global flows, nor to attempt to regress them into the Middle Ages of independent fiefdoms and princedoms. The indicated strategy is to channel globalizing trends into desirable pathways. Thus the second set of objectives focuses on creating a humanly manageable system of consultation and coordination in society—a strategy to promote the evolution of global holarchy.

Voluntarily concluded agreements among autonomous communities (we shall use the term "concords" to denote them) are proper instruments to achieve human control over global processes. Concords are needed in the economy, and in many other areas, such as science, art, religion, and culture in general. But they are urgent and imperative in two fields above all. The first of these is defense, and the second the environment.

Objective Number Three: Concords of Defense Cooperation

A few years ago, a Danish opposition party suggested that Denmark's entire defense budget should go into the making of a tape recording with the words "we surrender." If and when the country were attacked, the tape would be broadcast over national radio. The party lost the election—it had no chance of victory in any case—but its defense proposal struck a responsive chord. More and more people are coming to the realization that running up major defense bills to maintain a vast military apparatus is futile, especially for small countries such as Denmark. If such a country were attacked by a major power, its national army would be wiped out no matter how much money it had spent on it.

That national security calls for a powerful national defense force is a fiction; it derives from the illusion of the sovereignty of nation-states. If a country does not claim unconditional sovereignty over its territory it would have every reason to entrust the defense of its borders to joint peacekeeping forces. Such a step would make sense already today: the borders of a country such as Denmark would be more effectively safeguarded by a common European defense system than by a national army.

A joint European defense force, rejected by France when it was first proposed in the 1960s, is again in the making and may become reality before the end of the century. It would not take much persuasion to make the Danish people see the light; and there is growing support for it in countries such as the Netherlands, Belgium, Germany, Austria, and Italy. It would be more difficult to convince the majority of the English and the French to give up their national armed forces: their current national ethos includes a dream of major military status. But surprises cannot be ruled out, as the example of a popular referendum conducted in Switzerland in November of 1989 illustrates. In question was the Swiss army—a national institution long held in awe and respect by the whole population. Yet enough signatures could be collected by the Swiss socialists to force a referendum on whether to maintain the army or to abolish it altogether. The expectation was that not more than 5-6 percent of the population would say no to the army; but then came the surprise and the shock—over 30 percent of the ballots had the negative verdict.

Small countries in Europe may come relatively quickly to the insight that it is pointless for them to maintain an expensive army apparatus when with much smaller expenditures they could have a good internal police force and an adequate external security system. But even if Europe manages to evolve its own defense system, the US, China, and the Soviets do no appear ready to entrust their national defense to a—necessarily global—joint peacekeeping force like that of the United Nations. United Nations peacekeeping, though it has proved its effectiveness in Cyprus, in the Near East and in the Middle East—and was honored with the Nobel Peace Prize in 1988—remains limited to chronic trouble spots where the superpowers themselves are stymied. It seems that the greater the military establishment, the more difficult it is to integrate it into a collective peacekeeping system.

Effective regional and global peacekeeping may be utopian today, but it may not be so in the twenty-first century. If humanity decentralized modern nation-states into human-sized communities hallmarked by self-determination and autonomy, and without pretensions of sovereignty, the concords entered into by the new states and communities could cover issues of mutual security.

The fact is that it is logical for states that join together in the pursuit of economic interests to join together also in the sphere of mutual security. A region served by an economic community would be also well served by a defense union. There is, of course, always a danger that larger forces, even if they were created for defense, would ultimately turn aggressive. Wherever there is a concentration of power, even if the power is limited, there is a chance for corruption—and of dreams of grandeur. However, potential aggression in the world of 2020 could be averted by a universal ban on aggressive weapons and, together with the ban, the creation of a standing regional peacekeeping force joined with a regional peace forum.

In light of the experience of the Gulf crisis of 1991, evolving the modalities of regional peacekeeping is of the utmost relevance. Let us consider the matter in more detail.

Regional forces could be recruited from some already constituted national defense forces. Initially, the joint force could be kept small, embracing a standing militia of perhaps not more than fifty or a hundred thousand. Though modest in size, it must be properly equipped. It should have the capability of moving rapidly to any trouble spot in the region, and on arrival should have adequate weapons to deal with regional defense problems. This would not spell the need for major aggressive weapons, and certainly not for nuclear, chemical, or biological ones. If national forces were permitted to retain only such capabilities as are needed to maintain law and order in their communities, they would not pose a threat to regional peace and security.

In the event of internal or external aggression affecting a state, the joint peacekeeping force could intervene. The additional weight of that force, even if relatively modest, could tip the scales and bring about a cessation of hostilities. Once truce is established, the regional peacekeeping forum, constituted by the commanders of all national forces and of the joint force itself, could decide on measures to reestablish peace in the region.

The small, defensively armed system of regional defense could be bolstered by a series of mutual nonaggression pacts. Initial treaties could involve all regional forces on a given continent or subcontinent. Further treaties could be concluded subsequently among the parties to the various continental security treaties. With such a balanced and distributed system of global security there would be no need to call on the US or other major powers every time a regional conflict erupts, and humanity would have better chances of survival than by reliance on the precarious balance of terror which, so far, has managed to keep the nation-state system from holocaust.

The holarchic system of security would also enhance chances of prosperity. Local and regional economics would be freed from the burden of maintaining costly military establishments and could use their human and financial resources for productive ends. The advantages, as the current discussion of the "peace dividends" demonstrates, could be considerable.

Objective Number Four: Concords of Environmental Cooperation

The second area where concords would have to be rapidly and purposively created is the environment. To be sure, in this context just what we are to understand as "environment" has to be properly defined. It is more than birds and bees, flowers and trees, important as individual species are in themselves and as part of nature's diversity. By environment we must understand the biosphere as a whole, the total system in which man and nature are integral elements and interdependent partners.

Humanity, as other living species, can only survive in an environment where basic biosphere balances are properly maintained. Many of these balances have already been seriously damaged. We are on the way to a higher global heat balance; to some extent the greenhouse effect has become irreversible. We have thinned the ozone layer through our use of chlorofluorocarbons; and this, too, has passed the threshold of full reversibility. We have killed off countless species, and they can never be regenerated. We threaten a third of the planet's total land surface with desertification, and may have already sealed the fate of several tropical rain forests. Only time can tell the extent of the damage we have wrought, but it would be foolish to wait until time does tell. With each passing day the processes become one degree more difficult to turn around.

We do not know what point the degradation of the biosphere will have reached by the year 2020, but we do know the environment is already badly in need of protection. This is a global problem, and it must find a global response. In no area is "acting locally" as much in need of "thinking globally" as in the sphere of the environment. Obviously, global thinking must not remain abstract and theoretical; it must find concrete expression in globally coordinated action concorded by all states and communities. The holarchic world of the twenty-first century must protect nature from human shortsightedness as well as from human greed.

The objective of one set of global concords would be the regulation of mining and the use of natural resources. In today's world, sovereign nation-states proclaim themselves the absolute owners of forests, wetlands, croplands, rivers, and lakes, and of the metals, minerals and fuels found on the land and under the continental shelves of the seas. Not surprisingly, their use of these resources often does violence to nature. Roman law specified "*jus utendi et abutendi*"—that the right to use is at the same time the right to abuse. But if the communities of the next century would not claim unconditional sovereignty over their territories, they would not regard any part of the environment as their exclusive property. They would view all the environment as a precious resource handed to humanity on trust. The human right to *use* would not include the right to *abuse*.

Issues of the use and abuse of nature also concern territories that do not fall within the jurisdiction of any state or community. There are great quantities of industrially valuable metals and minerals under the continental shelves of the sea, and the arctic regions harbor additional valuable resources such as deposits of natural gas. None of these regions must be allowed to suffer irreversible degradation as a result of national or corporate greed and shortsightedness. If we were to accept the principle that we are the stewards of nature, we must accept stewardship whether nature is within the confines of our organized habitations or not. If we do not claim ownership over nature, we can regard all natural resources as a collective heritage, to be used for the joint benefit of the present and all future generations.

A second set of concords must aim at safeguarding the balances and regenerative cycles of nature from inadvertent intervention. Such concords would have at least the following goals:

- Setting and enforcing rigorous controls on the emission of chlorofluorocarbons (CFCs), the manmade gases that deplete the ozone layer and trap the Earth's heat.
- Setting and enforcing similar controls on the burning of coal, oil and gas, processes that give off carbon dioxide (CO_2) and add to the heat-retaining blanket around the planet.
- Setting upper limits on the use of the other trace gases (such as carbon and nitrous oxides, hydrocarbons and methane) that contribute to the greenhouse effect.
- Designing and implementing major reforestation programs to regenerate fuelwood and at the same time absorb CO_2.
- Designating up to 10 percent of the planet's land surface as protected areas for the on-site preservation of genetic resources.
- Carrying out soil conservation programs on impacted areas with attention to protecting the critical watersheds.

These concords would have to be complemented by a third set, having the creation and maintenance of environmental emergency capabilities as their objective. The specific goals would be the following:

- Identifying areas that are vulnerable to flooding if and when the polar icecaps begin to melt.
- Warning and if necessary relocating coastal populations threatened by irreversible increases in the level of the seas.
- Retraining and relocating farmers affected by changing weather patterns.
- Maintaining adequate rescue capabilities for use in the event of ecodisasters and ecocatastrophes.

Environmental concords would take money to implement. In today's world this seems like a major obstacle, but by the year 2020 financing may not be as difficult to come by. Adequate resources could be mustered first of all by recourse to the savings created by the reduction of military and defense budgets. Almost all environmental measures could be implemented in this way, especially in previously highly militarized countries. But ongoing financing would be required as well, long after the economies have adjusted from war to peace.

Long-term financing may have to take the form of environmental taxation. Its aim would be to maintain economic activity along environmentally sound and sustainable lines. Environmental taxes would not need to add to the total tax burden of a community; they could be compensated by the reduction of corporate and personal income taxes, and value-added taxes. The specific mix of taxes could be left to the discretion of individual states and communities; it would be enough that sufficient tax incentives be introduced to redirect economic activity from polluting and wasteful to clean and sustainable modes of operation.

For example, taxes could be introduced on the production and use of fossil and hazardous nuclear energies. This would have the effect of creating incentives for the research, development and application of renewable, clean and safe energies, and for enhancing energy efficiency. Taxes on polluted waste water and on clean water use above specific thresholds would induce water recycling, internal purification and water use efficiency. Similarly, taxes on unsorted solid waste, and proportionately higher taxes on hazardous waste would prompt improvements in product design, production processes, internal recycling and detoxification, as well as in technology choices and corresponding consumer preferences. The same would apply to the diffusion of chemical substances on land, in water, and in the atmosphere. Taxes could be levied even on urban and industrial sprawl, making it more efficient to clean up unused or derelict industrial and urban areas than to pave over virgin land. Through such incentives efficiency in energy and resource use would increase, and growth would be further shifted toward clean, information and labor-intensive activities such as services, crafts, arts, science, as well as human contact and communication in general.

Concords in the key areas of security, environment, politics and economics could lead to a global holarchy that is sensitive to human need, responsive to human purpose, and sustainable within the fragile and intricate generative cycles of the biosphere. As its values become enculturated, restraining measures could be phased out. Contacts among individuals and societies would grow and intensify spontaneously but they would not enslave. Humankind could look forward to a new era in civilization, with powers and capabilities conferred by integrated yet effectively mastered networks of information, and economic, social and cultural exchange.

The choice between evolution and devolution, and the further choice between evolution toward a global hierarchy, or a global holarchy, is still ours to make. Exercises in sharpening our "vision for 2020" develop our capacities of making it. Evidently, such exercises should not be limited to one man's views: they must involve many minds and many cultures. If they did, plans and ideas that today are but conceptual exercises could tomorrow become practical realities.

Even such miracles fall within the compass of possibility in the creative chaos of an age of bifurcation.

PART THREE

THE CRUCIAL NINETIES

CHAPTER 7

The Passing of the Modern Age

When an age is called the age of fear and uncertainty, when young people prefer not to think about the future—their own, or their children's—when most things one tries seem to bring unexpected side effects; and when the mass media revive the "good old times" as a sure way of pleasing the public, then something is seriously wrong with the times. It may be more than just a temporary abnormality, a painful but passing lapse after which everything will become sane and reliable again. It may be that the Modern Age is passing into history.

The Modern Age is the age that gave us industrial civilization, the nation-state, the automobile, television and telecommunication, and that extended human life expectancy from the medieval forty-odd years to over seventy. Its achievements stand undisputed. Its blessings, however, can be questioned. The technologies it created produce ever more unexpected interference with the delicate balances of nature, and they alienate, polarize, and threaten those they were supposed to serve. In the heat of its rapid industrial revolutions, the Modern Age seems to have overreached itself. The revolutions it brought forth moved from the sphere of technology and industry, into that of society and politics. The present and coming revolutions may bring more than a change within this age; they may spell the passage from this age into the next.

Culture and civilization are never passive, not even in epochal transitions. The great sociocultural systems of humanity do not just wait for their demise; they fight, struggle, and come up with innovation after innovation. Some are squashed by the outgoing age, but some can break through into the fresh space of the age that dawns. Social evolution, unlike a machine that can only execute what it has been programmed to do, has growth and momentum, flexibility and creativity. It has known many ages in the past and, with some luck helped along by timely insight, it will know many more in the future.

That an age is ending is not an unprecedented phenomenon. In the span of the last ten millennia there were many ages, each arriving as a breakthrough in the then dominant mode of living; each flowering as a seemingly eternal blueprint of human existence—and each passing into history, unmourned and sometimes even unnoticed, as conditions, values and institutions changed beyond its reach. This is what is happening to the Modern Age. Its benefits are undisputed, but they do not reach the majority of humankind. On the other hand its drawbacks affect everyone. For the three-quarters of humanity that lives in the underdeveloped world, the dream of material affluence through rapid modernization has failed. It has also failed for the people of the socialist countries, who now rise up to improve their lot. And those who enjoy the modern dream—the U.S., Europe, and Japan, and the newly industrialized countries of Asia—suffer from unexpected side effects: pollution, overcrowding, skyrocketing urban housing costs, mercurial trade restrictions, and unstable financial markets.

The problems of contemporary societies are neither accidental nor reversible: the Modern Age suffers from a terminal case of obsolescence. The conditions it has created, as outlined in Chapter 2, are basically unsustainable. Conditions, however, do not come about by themselves; they are created by societies with specific production and consumption structures, and people who hold values and beliefs that fit them into these structures and shape their future development.

THE OBSOLESCENCE OF MODERN BELIEFS

We should be able to tell whether the Modern Age is viable by looking at the typical values and beliefs of *Homo modernus*, the fabled, it at times almost fictional, prototype of the modern person. Let us take the following set of beliefs, beliefs that may be controversial, but which set off *Homo modernus* from all previous cultural types.

The law of the jungle. Life is a struggle for survival. Be aggressive or you perish.

A rising tide lifts all boats. If as a nation we prosper, all our citizens prosper and even other nations will do better.

The trickle down theory. Another watery metaphor, it holds that wealth is bound to "trickle down" from the rich to the poor, and the more wealth there is at the top the greater the trickle that reaches the bottom.

The invisible hand. Formulated by Adam Smith, it holds that individual and social interests are automatically harmonized. If I do well myself, I also benefit my community.

The self-regulating economy. If we could ensure perfect competition in a market system, benefits would be justly allocated by the system itself without need for intervention.

The cult of efficiency. We must get the maximum out of every person, every machine and every organization, regardless of what is produced, and whether or not it is needed.

The technological imperative. Anything that can be done ought to be done. If it can be made or performed it can be sold, and if it's sold it's good for you and for the economy. If nobody wants it, then you must create demand for it.

The newer the better. Anything that is new is better than (almost) anything that is last year's. If you cannot bring out a new product, call the old one "new and improved," and progress—and profits—will be yours.

The future is none of our business. We love our children, but why should we worry about the fate of the next generation? After all, what did the next generation do for us?

Economic rationality. The value of everything, including human beings, can be calculated in money. Everybody wants to get rich, the rest is idle conversation or simple pretense.

My country, right or wrong. We are sons and daughters of our great land, while all others are foreigners, out to get our wealth, power, and skills. We must be strong, to defend our national interests, preferably stronger than any possible adversary.

As these beliefs indicate, *Homo modernus* is a curious beast. He lives in a jungle, benefits mankind by his pursuit of material gain, trusts invisible forces to right wrongs, worships efficiency, is ready to make, sell and consume practically anything (especially if it is new), loves children but is indifferent to the fate of the next generation, dismisses things that do not have immediate payoffs or are not calculable in money, and is ready to go and fight for his country, because his country, too, must fight for survival in the international jungle.

Today, the beliefs of *Homo modernus* no longer pay off. Belief in the law of the jungle encourages tooth-and-claw competitiveness which fails to make use of the benefits of cooperation—especially crucial in a period of reduced growth opportunities and frequent squeezes. Holding to the dogmas of the rising tide, the trickle-down effect and the invisible hand promotes selfish behavior in the comforting—but sadly no longer warranted—belief that this is bound to benefit others. Faith in a perfectly self-regulating free-market system ignores the fact that in a laissez-faire situation those who hold the power and control the strings distort the operations of the market in their own favor, and push the less powerful and clever partners into bankruptcy. Efficiency, without regard to what is produced, by whom it is produced, and whom it will benefit leads to mounting unemployment, a catering to the demands of the rich without regard to the needs of the poor, and a polarization of society in the "modern" ("efficient") and the "traditional" ("inefficient") sectors.

The technological imperative becomes dangerous when economic growth curves slacken, markets become saturated, the environment approaches the limits of its pollution absorption capacity, and energy and material resources become scarce and expensive. Following the imperative issues in a plethora of goods that people only *think* they need; some of them they use at their own peril. That the newer would always be the better is simply not true: sometimes the newer is worse—more expensive, more wasteful, more damaging to the health, more polluting, more alienating or more stressful. One day a product is "improved" because it contains fluorocarbons, antihistamines, cyclamates, or just plain sugar—the next it is "improved" precisely because it does *not* contain these things. In the scramble for catching the public's fancy, health and social benefits are only pawns, to be used when they improve the marketing effort and ignored when they do not.

Living without conscious forward planning may have been fine in the days of rapid growth when the future could take care of itself, but it is not a responsible option at a time when delicate choices have to be made with profound and far-reaching consequences for future generations. If today we should shrug and say *après moi le deluge*, we would indeed bring about a flood—of overexploitation, overpopulation, inequality, and conflict.

The naive reduction of everything and everybody to economic value seemed rational in epochs when a great economic upswing turned all heads and pushed everything else into the background, but is foolhardy at a time when people are beginning to rediscover deep-rooted social and spiritual values and to cultivate lifestyles of voluntary simplicity.

And, finally, the simple chauvinistic assertion of "my country, right or wrong" can play untold havoc both domestically and internationally, calling for people to go and fight for causes which their country later repudiates, to espouse the values and worldviews of a small group of political leaders, and to ignore the growing cultural, social, and economic ties that now evolve among people in all parts of the globe.

The Modern Age is passing into history, but the values and beliefs of this age are still the basis of most of our economic, social and political practice. In the eighties *Homo modernus*, though far from well, was alive and kicking. But whether he can survive the crucial nineties is open to question.

Table 1
The Changing World-View in the Age of Bifurcation.

	The Modern View	The Emerging World-View
Physical world	Atomistic; fragmented. Objects are independent and free-standing. People are individuated and discrete.	Holistic; interconnected. Objects and people are interwoven into a community.
Physical processes	Materialistic; deterministic, mechanistic.	Organic; interactive, holistic
Organic function	Discrete and separable; parts are exchangeable.	Interwoven; interdependent. Not interchangeable or exchangeable.
Social ethos	Technology oriented; interventionist; goods-based.	Communication oriented; service based.
Social progress	Consumption dependent; resource conversion.	Adaptation oriented; balance of resources.
Economics	Competition and profit driven; exploitative.	Cooperative and information driven.
Humankind	Mastery over nature. Anthropocentric.	Integrated into nature. Gaia-centric.
Culture	Euro-centric; colonial.	Pluralistic.
Politics	Hierarchical; power-based.	Holarchic; harmony-based.

CHAPTER 8

The Holistic Alliance

In the early nineteenth century, after the French Revolution and the Napoleonic wars, Europeans forged a "Holy Alliance" dedicated to the creation of a community of the Christian nations of the world. All nations were invited to the alliance as long as they professed Christianity, regardless of their role or fate in the foregoing wars. While ultimately the Holy Alliance fell apart, during its reign it brought about a system of collective security with lasting and far-reaching benefits.

There is something to be learned from this. The strength of will and motivation we need today is similar to that which sparked the Holy Alliance. But the alliance we need in the ultimate decade of this century—and of this age—must be *holistic* rather than *holy*. The holistic alliance is to link new and progressive currents in science, art, religion, and education in facing the common challenge of the coming bifurcation. This is a major societal mission, similar to the Apollo mission that in 1969 landed the first astronauts on the moon. The holistic alliance must be dedicated not only to landing a few men on the moon, but to all men and women in the next age, here on Earth.

Is this a feasible, even a desirable mission? Can one, should one, commandeer science and art, religion and education? This may be equivalent to custom-designing cultural change—and the historical precedents for it are not encouraging. In the nineteenth century, Marx and Engels wanted to use science to change the capitalist culture of Germany and England. If they have succeeded in provoking changes in twentieth century Russia it was because Lenin used their theories to underpin the power politics through which he managed to dominate a rebellious and war-torn tsardom. Stalin wanted to use "scientific socialism" to eliminate the remnants of bourgeois culture in the Soviet Union and he, too, failed despite his ruthlessness and the vastness of his propaganda machinery. Mao hoped to use the philosophy of his "little red book" to wipe out all traces of the traditional culture of China and, even though his Red Guards were brainwashed and violent, he ended in failure as well. Statesmen and dictators, benevolent and malevolent, have long recognized the power of art, science, education, and religion to change the way people think and act. None have succeeded. Should this not teach us a lesson about similar attempts in the future?

Clearly it should. But the difficulty of creating purposive change in culture only compounds the problem. After all, what we need is a fundamental change in the way we think about ourselves, our environment, our societies, and our future. If we change our policies and our technologies without changing ourselves, we do not evolve our condition—we only produce temporary fixes without enduring effects. Only a basic change in the values and beliefs that guide our thinking and acting is of lasting consequence. Such a change, however, amounts to a transformation of culture—to a leap in "cultural evolution."

Fortunately, cultural *evolution*, unlike the Maoist brand of cultural *revolution*, need not be forced "from above." It can come to the fore in indigenous developments in science, art, religion, and education, and permeate from the new consciousness of the public to the thinking of men and women in positions of leadership. The fact to recall is that in an epoch of bifurcation all the structures of society become highly sensitive: they register and change with every minute fluctuation. Culture is no exception. In the transitional epoch new ideas and values, initially small and powerless fluctuations, surface in profusion. Once surfaced, some among them are bound to "nucleate"—get hold of the imagination of wide layers of the population and change dominant modes of thinking and behaving. If also ideas and values that inspire historically adapted and humanly beneficial trends would surface, they, too, would spread in society and influence cultural change. To trigger cultural evolution the heavy hand of dictators is not needed; it is enough to create well-conceived fluctuations in values and beliefs, and support them in the welter of competing ideas and movements.

It would be a tragic mistake to interpret the challenge of the age of bifurcation as a call to use science, art, religion, and education to achieve a preconceived end. The response to the challenge can be more modest; it can rely on spontaneous cultural evolution. It can demand that scientists, artists, religious leaders, and educators cultivate their social consciousness. This is by no means unreasonable and it does not call for coercion. It calls for a revitalization and resurgence of the sense of responsibility that already emerges among responsible people in the arts, in the sciences, in education and in religion, and for the comprehension and support of the contemporary leadership in allowing the new motivations to surface and to achieve effective social impact. If the main carriers of culture would act as responsible agents of cultural evolution, the outcome would no longer be left to chance. While it would be supported "from above," it would not be imposed: its motivation and inspiration would come "from below"—from within the structures of contemporary culture itself.

THE DEMAND ON SCIENCE

The aloofness and introversion of contemporary scientific communities is not due to a quirk in the personality of scientists: it has long and deep historical roots. These go back to the origins of modern science in the sixteenth and seventeenth centuries. It was at that time that the humanistic culture of Europe extricated itself from the domination of the medieval Church. At first, the influence of religious precepts was so strong that the initial orientation of scientific thinking became colored in reaction to it. Science was to be impartial and disinterested; it was not to intrude on the sacred authority of the Pope. The trials of Giordano Bruno and of Galileo gave ample proof of the power of the medieval spirit over scientific inquiry. The young sciences could grow only by abstaining from interference in the affairs of society and professing both independence and disinterest.

The latter viewpoint proved to be patently false. Science became one of the greatest forces molding modern civilization, far greater than the religious influences from which it first hoped to retreat into neutrality.

In the years since World War II, ever more scientific theories were translated into practical technologies. The impact of the natural sciences was equalled by certain branches of the social disciplines, especially economics. Far from being a search for truth *sub specie eternitatis*, science proved to be a crucially important social, political and economic activity.

In our day, the idea of scientific neutrality and disinterestedness must be relegated to history. This does not mean a surrender of scientific objectivity, only a recognition of its limits. As long as scientists remain dependent on society for pursuing research, they will be influenced by social priorities. And as long as they concern themselves with matters that have applications to, or even just implications for, human beings and society, they will be witting or unwitting agents of cultural change.

Throughout the 1990s there will be a great need for relevant scientific knowledge; many vital questions for scientists to consider. Will we be able to control the forces which, if left unchecked, would lead to global crisis and perhaps mass destruction? Will we be able to create and sustain a global holarchy in which no state, no one society is in control? Can people interact and communicate without inducing dependence on each other—especially dependence of the weaker and more naive (or honest) on the stronger and less scrupulous? Can there be effective limits to growth—the growth of population, of cities, of power and of wealth? Can technology be controlled and made to serve human needs and objectives instead of becoming an end in itself and creating its own needs and demands? Is there a way to satisfy people's needs for privacy and personal space despite high levels of communication and large numbers of people sharing the same physically limited planet? Can this planet support 10 billion people or more without irreversible damage to its ecology?

Most crucial of all, can people share the planet with tolerance and mutual respect? The society of the future is bound to be diverse and pluralistic; it could also be decentralized and grassroots oriented. This means a holarchic system which has local autonomy as well as global coordination. To understand how such a system could work, one has to model it. But the needed models will be different from the dominant social systems models of the twentieth century: these were inspired by a single culture—the Western—and assumed individual and institutional behaviors based on a single type of rationality—likewise the Western.

The demands on scientists are great, and they are distributed throughout the social and the natural sciences. These are not problems for the sociologist or the political scientist alone. They are also problems for the ecologist, the urbanist, the psychologist, the demographer, the economist, the chemist and the physicist—and for the cybernetician and systems scientist. Within the current boundaries of the disciplines, no scientist is able to successfully confront them.

Fortunately, disciplinary boundaries are not eternal. They are a heritage of the past, and now they have become outdated. Each of the great new fields of scientific inquiry—the new physics, the new biology, and the new systems sciences—seeks and finds major strands of unity in the world's manifest diversity. The findings cohere into a remarkable picture of the world, one where the universe is self-organizing toward progressively higher levels of evolution, with complexity balanced by integration. The current scientific revolution is as great as that which replaced the Copernican Earth-centered universe with the modern concept of the solar system, and it is far richer in human and social implications. As this revolution unfolds, science gains greater and greater social relevance. It is not any the less good science for being relevant to human concerns: practical usefulness and sound knowledge do not exclude each other. Integral theories of nature and society make good science, as well as reliable sources of useful information.

THE DEMAND ON ART

Artists were the principal architects of the Renaissance and their human and social relevance has not diminished in our time. In this age of bifurcation the social responsibility of artists is as great as that of scientists. But, as we have already noted, contemporary artists have grown even more aloof than scientists from the concerns of society.

The separation of art and society, unlike the disinterestedness of science, is a typically twentieth century phenomenon. Until then, most artists were integrated individuals; they had inseparable human, social, political and artistic interests. From Aristophanes to Balzac, writers stressed this unity of concern; Picasso's "Guernica" gave an eloquent demonstration of it. Plato's assertion that truth can also be apprehended as beauty was echoed by Schiller, who in this poem *The Artists*, said, "What we have here perceived as beauty, we shall some day encounter as truth." Balzac claimed to complete with the pen what Napoleon began with the sword, and figures like Goethe and Wagner did not hesitate to convey a social and cultural message through their works.

In literature, such writers as Herman Hesse, Jean-Paul Sartre and Eugene Ionesco remained dedicated to this tradition, but many of the arts managed to divorce themselves from concern with society at large. Music, painting, sculpture, even dance, turned ever more inward, in search of "internal" laws and meanings. It was Arnold Schönberg who scoffed most explicitly at the notion that art has to address society at large. If it is art, he said, it is not for all, and if it is for all, it is not art. Many composers of today's avant-garde share this sentiment. The music of a Stockhausen or Boulez cannot be understood by the layman; as one devotee said, one does not know if a piece is good until one has analyzed the score. The same can be said of the majority of the painters and sculptors whose works are hung in prestigious galleries, and whose names are sacred to the "in" circles of the art elites.

It is getting more and more difficult for an artist to be considered very good and be at the same time very popular. In 1913 Stravinsky's *Rites of Spring* caused a furor; in 1991 the first performance of an avant-garde work elicited the interest of but a few critics and the awed approbation of only a small and inbred coterie of followers. Works of art are traded for their prestige value, or as investment. People visit galleries, museums, concerts and the opera for irrelevant reasons—it is part of one's education, or the socially proper thing to do.

The majority of twentieth century artists reject society as their public, and society has all but given up on "high" or "serious" art as a source of enjoyment. Of course, great art was never enjoyed by all the people, even in the eighteenth and nineteenth centuries. But in those times such art was reserved for royal and princely courts, and the nobility in their entourage. In our day a much vaster public would be ready to enjoy art of all kinds, were it not for the introversion of leading artists, critics, and historians.

Such attitudes must not be prolonged in an epoch in which the creative minds of society need to focus on vital choices and unique opportunities. After all, great art disciplines the imagination, leads to fresh insights into human nature and the nature of social relationships, and provides guidance in the selection of goals and ambitions.

There is, however, a vexing problem that arises as soon as art is called upon to serve social causes. Does this not infringe on the freedom and autonomy of art? Would it not interfere with the primary concern of artists, which is their own self-expression through their chosen medium?

Few have contested the horror of the forced enslavement of art by politics in the 1950s. "Socialist realism," as Stalinist art was known, produced poor propaganda and even poorer art. Revolutionary art, whether in China, Africa, or in Latin America, likewise shortchanged artistic integrity in favor of promoting political causes. Art in the service of politics is as bad as art in the service of profit. Even if an occasional brilliant poster appears on the streets of Vienna, London, Paris or Berlin, on the whole commercial art is little more than illustration. Should artists not keep away from all social causes and protect the integrity of their art and creativity?

In the crucial epoch of the 1990s this question takes on particular urgency. The excesses of political art have to be avoided. But that does not mean that art should be divorced from society and absolved from social responsibility. Even as self-expression, art is addressed not only to the artist himself but to a broader public: the communication of the esthetic experience is an integral part of artistic creativity.

Artists could be free, spontaneous and creative, yet socially relevant. Their works could teach the eyes to see, the ears to hear, and the mind to comprehend human reality in the many splendors of its evolution. Great artists familiarize the new and humanize the vaguely threatening, crystallize half-thought-out notions and give birth to a wide array of values and ideals. The performing arts in particular, have a wide and deep social impact. Theater, television, and motion picture can galvanize adoration as well as trigger controversy. Dramatic works on stage and on the screen launch new trends, not only in the arts but in society at large.

Regaining their social relevance, art and science, the twin expressions of the "high" culture of contemporary civilization, must become major engines of its evolution.

THE DEMAND ON RELIGION

Belief systems do not become superfluous, even when science and art are tuned to human concerns. Science does not address issues of ultimate meaning and truth, not to mention divine will and purpose. Art does occasionally embark on themes of transcendental significance, but it treats them in an esthetic and intuitive, not in an explicit and systematic way. In any case, there is more to human beings than scientific reason and esthetic sensibility—there is also a spiritual dimension that neither science nor art can fully satisfy. Religion is there to respond to this need.

The great religions have not only provided the means for individual fulfillment; they also gave guidance to harmonize social relations. The social and ecumenical element in the religious tradition is evident in the Judeo-Christian faith no less than in the belief systems of the Orient. Judaism, for example, sees man as God's partner in the ongoing work of creation and calls on the people of Israel to be "a light to the nations." At the heart of the Christian teaching is love for a universal God that must be reflected in love for one's fellows and service to one's neighbor. Although Moslem fundamentalists keep fighting holy wars against heathens, Islam, too, has a universal and ecumenical aspect. *Tawhid*, the affirmation of unity, means the religious witness "there is no god but Allah"—and Allah is the symbol of divine presence and revelation for all people. Hinduism, unique among the great religions in not having an individual founder, perceives the essential oneness of mankind within the oneness of the universe, and Buddhism has as its central tenet the interrelatedness of all things in "dependent co-origination," interpreted by progressive Buddhists as a mandate for achieving higher forms of unity in today's world of interdependence.

The Chinese spiritual traditions revere harmony as a supreme principle of nature and society. In Confucianism harmony applies to human relationships in ethical terms, while in Taoism harmony is an almost esthetic concept defining nature, and the relationship between man and nature. The Baha'i faith, the newest and most rapidly growing of the world religions, sees the whole of mankind as an organic unit in the process of evolution toward peace and unity—a condition that it deems both desirable and inevitable.

These are significant elements in the world religions but, with few exceptions, they do not come sufficiently to the fore. They are overshadowed by parochial concerns, and the competition between particular faiths offering a unique path to fulfillment and salvation in exclusive possession of the truth. A new emphasis on the ecumenical, more valuable side of the coin would not do violence to the doctrines; it would only make them more relevant. The leaders and prophets of the great religions have claimed to be humanly and socially relevant to their time; their followers could not object to their teachings remaining relevant to *our* time.

If contemporary religions were to become truly relevant today, they would need not only to recover the humanism of their traditions but also to forge ahead to give new meaning to life in our age of bifurcation. To achieve this objective a return to fundamentals, no matter how enlightened, is not enough. There must be a new development, a creative extension of the ideas that have informed and inspired the great religions since the dawn of civilization.

The basic contours of this development can be already discerned. To begin with, one of the basic concepts of the Judeo-Christian tradition must be abandoned: that of a God external to man and the universe. The God of contemporary humankind must be an immanent God, inspiring the world from within, not commanding it from above. This concept is not foreign: in Christianity it appears in the naturalism of St. Francis of Assisi as well as in the evolutionism of of Pierre Teilhard de Chardin. Then, the self-creation of the universe must be recognized, and indeed celebrated. This is more of a challenge: the biblical tradition has not come to terms yet with an evolving reality. Most Judeo-Christian religions, although they have an historical perspective when it comes to the spiritual development of the individual, do not have a corresponding perspective on the evolution of humanity. Yet such evolution occurred, and it took place within the larger context of the evolution of life on Earth, and of the Earth in the cosmos.

For the most part, contemporary theologies perceive a divine kingdom reigning in an established and unchanging universe. The antagonism of some streams within Christianity to the concept of evolution is only the surface manifestation of the abiding unease with which the Judeo-Christian tradition faces the reality of fundamental change taking place in the very fabric of the world. In our day such change can no longer be neglected: we are now in the midst of a basic and irreversible process of transformation that affects all life on this planet. To remain relevant, the Western religions would have to convey a view not of an abiding and perhaps seasonally renewing world, but of a fundamentally and irreversibly evolving one.

It has been the historical task of the great religions to perceive and proclaim the spiritual aspect of the world, and to acquaint the faithful with its meaning. In the next development of religion, the perception of the spiritual aspect could lead to the image of a self-ordering cosmos. Theologians could recall that we are ourselves children of the evolving universe; that we bear within ourselves the impress of every transformation through which the universe has ever passed. The elements of which our bodies are composed have been created in the fiery processes of stellar interiors and stellar supernova bursts. They passed through a phase of dispersion in interstellar space, to be brought together in the womb of the protostars of a new stellar generation. As elements on the surface of the planets born of these stars, they have participated in the original emergence of life in rich mixtures of molecules and protobionts in primeval seas. They entered and left living bodies for billions of years, cycling through the web of structured connections that make up the self-maintaining and self-evolving reality of the Earth's biosphere. The forces that brought forth quarks and photons in the early moments of the radiance-filled cosmos, that condensed galaxies and stars in expanding spacetime, and that created complex molecules and systems on life-bearing planets—these forces are at work in our own bodies. They inform our brains, infuse our minds, and come to self-recognition as we gather the many strands of our new knowledge of the universe.

The new evolutionary knowledge, elaborated in science, could be deepened and made humanly meaningful in religion. Religious communities could celebrate the original flaming forth that gave birth to the known universe—the sudden synthesis of photons and the many microparticles, and of atoms and molecules throughout the expanding reaches of cosmic space. They could celebrate the emergence of macromolecules and protocells, the precursors and harbingers of life, on the surface of this planet as well as countless as yet unknown planets in this and myriad other galaxies. They could recognize that the cosmos is our larger self; that our journey as individuals reflects the epochal journey of the world. They could come to know the embracing, irreversibly evolving universe as our primary sacred community.

No creation story could be more inspiring than this concept; none could reflect as well the reality of the physical aspect of reality. And none could be more timely. Because, by recognizing and celebrating the self-creation of the world, religious communities would recover an ancient insight: they would again regard nature as sacred—an element in the sacred community of the cosmos. With the recovery of the sanctity of nature, contemporary people would receive fresh inspiration to reorient their attitudes toward the natural environment. Religious communities would re-sanctify nature and participate in humankind's emergent evolution with full spiritual force.

Religious renewal has always come in the wake of civilizational crises. It was in the disastrous moments of the history of Israel that the prophets of Judea made their appearance; Christianity established itself in the chaos left by the moral weakening of the declining Roman Empire. The Buddha appeared in a period of spiritual and social confusion in India; Mohammed proclaimed his mission in an epoch of disorder in Arabia; and the religion of Baha'ullah emerged in confinement imposed by a moribund and strife-torn Ottoman Empire. Today, we are in the throes of the greatest and deepest crisis our species has ever known, in an epoch when the very web of life on Earth is threatened. Will there be another great renewal in the womb of the age of bifurcation? If this should come to pass, the religious communities of the world could enter the holistic alliance as a vital cultural force, illuminating and encouraging the next step in our species' evolutionary self-creation.

THE DEMAND ON EDUCATION

Science, art, and religion could be effective in meeting the challenge of the crucial nineties only if the ideas, intuitions and convictions generated by them can spread in society. Although science, art, and religion influence the thinking and feeling of practically all people, individuals who are neither science- nor art-minded, nor yet religious in any conscious way, would be slow to respond to the emerging views and visions. Such people—the bulk of the population in many lands—would have to be reached by more direct and popular means. This suggests, first of all, a different role for the popular media.

The commercial mass media—newspapers, radio and television among others—could be highly effective in spreading the relevant message, but it is difficult to see how they could transform quickly and effectively. Their current orientation to "timeliness" (meaning short time-horizons), and "human interest" (meaning mostly local relevance), would change only if the interests and demands of the public did. That, however, is likely to take time—time for new ideas to percolate and make their impact. This makes for a chicken-and-egg situation: the popular media will not change before the public changes, and the public will not change until the popular media offers the relevant kinds of information. The commercial mass media alone is not capable of breaking this vicious cycle. On the other hand the non-commercial public media could perhaps do so.

If public educational media are to meet this challenge, they must undergo major reform. In Western democracies they have better chances in this regard than the commercial mass media: public radio and television and the related educational outlets are not directly dependent on the fickle currents of popular taste. In such media journalists could exercise the responsibility they so like to evoke: they could raise issues that go beyond short-term timeliness and local human-interest concepts to embrace long-term issues of worldwide relevance. They would not need to couch such issues in the esoteric jargon of the sciences, not even in the often pompous packaging of scientific documentaries. Drama, comedy, poetry, and an imaginative multimedia treatment of the current issues and the opportunities they offer for the future would have significant positive impact.

The effort of the public media would be important, but in itself it would not be sufficient—the institutions of the cotemporary educational system would have to follow suit. There are opportunities for the transformation of all branches of the educational system, but they are not free from difficulties. Today's institutions are impregnated with outmoded concepts of the world, and of one's place in the world. They are fragmented along the fault lines of the natural-scientific-technical, the social-scientific-political, and the artistic-spiritual-religious subcultures. These divisions—the same as those between the hard sciences and the humanities—have become both obsolete and dangerous. They prevent people from acquiring an integrated vision of themselves and their age; from seeing things in the required integral perspective.

A reform of the educational system in the domain of the social sciences is of particular importance. Social and civic study programs in almost every part of the world foster what boards of education euphemistically call the "national ethos"—an ethos that in reality is often at the root of adult ethnocentrism, narrow in-group loyalties, and outright chauvinism. Such programs may end up by fueling international and intercultural misunderstanding and intolerance. As children grow up, the categorizations they learn in school become internalized and part of their personality. They are expressed in attitudes that influence social and political processes, not only in their country but, through the posture of their country, in the rest of this interdependent world. It is imperative that schools should no longer inculcate a narrow and short-sighted ethos, and that textbooks should cease acting as chauvinistic filters of national and world affairs.

This calls for many changes. Systematic analyses of the textbooks used in civics and social studies courses in US and European schools show that the usual emphasis is on the country's own history rather than on the history of others; that events and episodes from the history of one's land are presented in a manner that encourages children to believe that their country is superior to foreign countries; and that when foreign countries are presented they appear either as "friends" or as "enemies" rather than in terms of their own values and achievements. The texts seldom call for debate and critical thinking; mostly they call only for acceptance. Primary and secondary school teachers seldom spur debate; they do not like to deal openly with political controversy. Their role, they mostly believe, is to help children become loyal citizens with due respect for, and obedience to, public and institutional authority. As a result the modern educational system promotes conformity, passivity, parochial sentiments, and narrow and short-term outlooks. This state of affairs is as obsolete as the segmentation of the knowledge system is in the sciences. There is no contradiction between international solidarity and loyalty to one's nation, between being a good member of a family, a community, a profession and a nation, and being a good member of the world community. Indeed, the opposite may be true: a recognition of the legitimacy and value of other nations and cultures may be a precondition of perceiving the true fit of oneself and one's country in the global family of all people and all countries.

On the whole, the more prestigious the school, the more it reflects and inculcates the views and values of the society that puts it on a high pedestal.

This is a major stumbling block to the reformation of "prep" schools and elite universities. In such august institutions tradition tends to be valued to the extent that it restricts vision and filters innovation. It is not surprising that such institutions tend to produce leaders narrowly dedicated to the status quo. What our world needs, however, is not status quo-oriented leaders emerging from the ivory towers of specialized scholarship, but flexible and functional learning environments where people, young and old, can be exposed to concepts and ideas relevant to their present and to their future.

Flexible, task-oriented educational systems do not obviate the need for high-quality education, in schools where the finest fruits of knowledge are cultivated and communicated from generation to generation. But tradition-minded institutions, no matter how expert in specialized fields of scholarship, have to be complemented by institutions where knowledge is wide-ranging, integrated, and whole. The world of the late twentieth century needs a brand of institution where all people can gain an overview of the problems that beset their age. These, too, must be institutions of excellence, but they must be committed not to specialization but to the integration of knowledge and the development of a holistic vision.

Holistically oriented institutions need not emulate the pedagogy practiced in institutions of specialized learning. The overview needed in our crucial epoch is best acquired outside formal lecture halls. Learning could take place in the framework of informal seminars, debating and discussion groups, individual study under the guidance of tutors, and learning-by-doing internships. Holistic learning is a collaborative elaboration of insight, knowledge, and skill. Young men and women not intending to enter a specialized technical field need not move to sheltered campuses, nor should older people intent on gaining a better understanding of contemporary problems re-enter formal classrooms.

Holistic education is not a kindergarten, soon to be left behind by the smart and the ambitious. Rather, it is an essential propaedeutic to all forms of learning, even to those that are rigorous and specialized. The overview we all require is not a *simplification* but an *integration* of the latest fruits of contemporary knowledge. Such integration is an ongoing task, and it must have a home base. We need institutions that convene groups of "specialized generalists" to assess and integrate the emerging insights of the contemporary sciences, as well as those of the arts and the major belief systems.

It may appear that the task is too ambitious; that it calls for creating a new "renaissance man," able to master the full range of the knowledge of his time.

In reality the requirement is more modest. Given the current revolution in the sciences, achieving a holistic overview does not call for individual genius. The task can be systematically attempted; indeed, there are a few institutions that attempt it already.*

Centers for holistic learning have a vital role in the crucial nineties. In the absence of the integration provided by such specialized generalists, the alliance between science, art and religion would remain external and superficial. There must be institutions that provide a forum for informed individuals who make it their life's ambition to follow current developments in the sciences, the arts, and the religions, and to integrate them in consistent views of man, nature and society. This means doing more than putting concepts and theories side by side or one after another, as in a dictionary or an encyclopedia; it means showing how insights and theories cohere into a complete organic whole. Individuals of this kind are surfacing now in a wide variety of areas, from astrophysics to corporate management. When brought together, their emerging insights can cross-fertilize each other and receive mutual reinforcement.

Through the holistic alliance of science, art, religion, and education, we could learn to view the problems and challenges of this age of bifurcation as elements in a complex but unitary historical process. Our times of transition would be seen from mutually complementary angles. From the angle of the sciences, it would be perceived as the evolution of a specific variety of complex system, in which the universal laws of systems development take on the forms and characteristics proper to contemporary humanity in its global milieu. From the perspective of the arts, it would be seen as an adventure replete with drama and deep significance, offering fresh opportunities for personal creativity and for creating new relationships in living, loving, companionship, and solidarity. In the optic of the religions, the transition would appear endowed with still deeper meaning, as the emergence of a higher step in the ongoing self-evolution and self-integration of the universe, coming to ex-

---

*One such institution is the Vienna International Academy, a newly established private university that this writer serves as Rector. The stated aim of the academy is to promote holistic thinking, based on integration of leading-edge knowledge in the natural and social sciences and a creative dialogue between the sciences and the humanities. Its members, drawn from the sciences as well as the humanities, from government as well as the business community, share the ambition of becoming genuine generalists, and enabling others to achieve similar competence.

pression in new forms of consciousness that signify progress along humankind's long road toward oneness and unity.

The holistic alliance—the cultural Apollo mission of our times—can be forged. Achieving it calls for the growth of a new dimension in the social consciousness, sense of responsibility and effectiveness of scientists, artists, educators, and men and women of religion. It calls for the comprehension and support of business and government, and for the creation of flexible learning-researching institutions where groups of specialized generalists can integrate their emerging insights into usable foresights.

CHAPTER 9

The Ultimate Question

THE GAMBLE

Modern man no longer relies on the muscle of fight, the speed of flight, or the protective mask of shape and coloring for survival. He has come to depend on intelligence for his life. This is a fateful gamble. It has put at stake our collective survival, and that of the whole biosphere.

About five million years ago the evolutionary line that led to modern man diverged from African apes, the common ancestors of humans, chimpanzees, and gorillas. Apes are knuckle-walking quadrupeds; Homo is an erect biped. Apes have large jaws and they have small brains (in the range of 300–600 cubic centimeters), Homo has a small jaw, and a four-fold brain size in the range of 1400–1600 ccm. Most apes are adapted to life in the trees; Homo is suited to life on the ground. It is this adaptability to terrestrial life that proved to be the decisive factor in the evolution of intelligence. Why some bands of pre-hominids left the trees is still somewhat mysterious (some anthropologists maintain that they were pushed from the forests into the savannah by physically more developed arboreal primates) but once they left the trees, their destiny was sealed: they were condemned to a form of intelligence—or to extinction. The question we now face is whether the kind of intelligence that evolved is sufficient for survival into the twenty-first century. Humanity, as Buckminster Fuller said, is facing its final exam. It is an exam of intelligence: the collective IQ test of the species.

Intelligence in a species is not unique to Homo—other animals have developed forms of it, and more species might have developed it had they the need and the opportunity of doing so. Wales and dolphins have intelligence, but they live in an aquatic environment that is more stable and more suited to living beings than life on land. The intelligence of sea mammals had no need to evolve into the kind of active, manipulative intelligence of land-living humans. This kind of intelligence is needed only in a terrestrial setting, where the availability and retention of water, the ongoing procurement of free energies, and the maintenance of constant temperatures are essential to the running of complex biochemical reactions. A corresponding kind of intelligence may have emerged in various land-living species; in time it may have emerged among the dinosaurs. One species, the stenanicosaurus, had favorable prerequisites—a spacious cranium, large eyes and long arms—but it disappeared along with the rest. Had stenanicosaurus evolved with a high level of intelligence, the biosphere may now be populated by reptilian rather than human beings, with mind-boggling consequences.

Unlike the history of dinosaurs and of sea-living mammals, the chance concatenation of circumstances that made up the history of our own species allowed, and even required, our ancestors to stake survival on a manipulative form of intelligence. The gamble had to be taken because, once they were out of the trees, our forebears found themselves in a perilous situation. The savannahs were already populated with meat-eating animals, most of them stronger and faster than they. The shelter of the trees was gone, and in its place they had only one substitute: their newly freed forelimbs. These were no longer needed to hold on to the branches of trees and could thus be put to other uses. Most probably, the evolving arms were used to transport infants as the bands of early hominids followed migrating herds on Africa's developing grasslands. But they must also have been used for self-defense with stones and sticks, as chimpanzees use their forelimbs. Unlike in chimps and other apes, however, our forebears' method of survival put a premium on bodily control, tactile sensitivity, and especially on manual dexterity. Only those bands of hominids could survive that evolved these capabilities. Our early ancestors managed this feat: in the motor and sensory cortex of the brain of sapiens, the representation of the hand, especially the thumb, became phenomenally detailed.

As forelimbs transformed into dexterous arms and hands, jaws were no longer required for defense. There was no selection pressure for canine teeth, sectorial premolars and a capacious jaw to accommodate them. The pressure was for a bigger brain capable of dexterity and intelligence, and for a cranium to shelter it. Hence an erect bipedal species arose, with large brain, small jaw and counterposed thumbs—the hallmarks of sapients to this day.

With the development of a larger brain came a whole series of evolutionary innovations. Among the abilities that were advantageous to terrestrial bipeds the ability to cooperate in performing the critical tasks of survival may have been the foremost. Mutant individuals who had a superior ability to communicate with each other were favored by natural selection. As these socialized individuals diffused, the genetically based sign language of the apes transformed into the flexible system of shared symbols characteristic of human language. Social behavior was freed from the rigidity of genetic programming and became adaptable to changing circumstances. In the neocortex the capacities for manual dexterity and tool use were joined with newly evolved capacities for communication and socialization. Our forebears evolved from terrestrial apes into a species that, with some exaggeration but not entirely without reason, came to view itself as "the knower."

Sapiens remained essentially unchanged since they first emerged in Africa about 100,000 years ago. But their sophisticated manual and cognitive capacities did not pay off during the greater part of the 5 million years since Homo first descended from the trees. During most of these thousands of millennia the scattered bands of hominids just scraped by, barely holding on in an existence that was always vulnerable and frequently precarious. The payoffs began slowly, perhaps 1.5 million years ago. Near Chesowanja in Kenya archaeologists have found baked clay next to hominid bones and manmade stone implements. The clay showed traces of exposure to heat much higher than that which would normally occur in a bush fire. Whether it had been baked by fires tended by hominids who lived 1.5 million years ago is uncertain; the evidence is circumstantial—natural fires leading to an intense smoldering of a big tree trunk could have produced similarly high temperatures. But 500,000 years back in time, the evidence becomes uncontroversial. Fires of human origin are at least as old as that—and so are the first indications that our species' gamble on intelligence would eventually pay off.

The control of fire was an intelligent move: it gave the dispersed bands of hominids a small but decisive edge in their struggle for survival. Fire inspires fear in all creatures—flames and embers burn feathers, fur, hair, and skin on contact. Since the instinctive reaction is to flee, those who master fire can use it for protection and defense. Fire is also an important aid in assuring a continuous food supply; meat that quickly rots when raw remains edible when properly roasted. By roasting the food, lean periods between hunts in poor weather can be bridged; one is no longer living entirely from hand to mouth.

Mastering fire, the most immediate and fearful of all elementary forces of nature, is not likely to have come about all at once, and in one place only. *Homo erectus*, our direct forebear, seems to have tended fires in far distant locations over long periods of time. The finds speak clearly: there were humanly laid fires at such diverse sites as Zhoukoudian near Beijing, Aragon in the south of France, and Vértesszöllös in Hungary. A number of hominid bands seem to have arrived at the insight almost simultaneously, without learning from, or probably even knowing of, each other.

The process must have been slow, at least by modern standards. There are fires ignited periodically by lightning in all tropical and subtropical ecosystems. Natural fires play a vital role: they clear away dead organic matter and revitalize the soil, creating favorable conditions for fresh plant growth. *Homo erectus* will have encountered natural fires for untold millennia and will have reacted much the same as other apes and animals—by fleeing. But gradually, some adventuresome individuals were drawn back to the smoldering remains and began to poke around in them. No doubt, they discovered the remains of many kinds of animals, and found some that were charred but not entirely burned. Experience might have shown them that such remains could be eaten, not only at the site of the fire, but at home-bases for days afterward.

More and more of the exploring hominid bands will have returned to the sites of natural fires to forage for edible remains. They would not have been the only ones to do so: other animals, especially the readily imitating apes and monkeys, would have followed suit. But hominids had an advantage: with their thinly haired bodies they were less likely to be singed by flying sparks than more furry or hairy animals. Their erect posture was even more of an advantage. Liberated arms could be used far better to investigate embers and ashes than the forelimbs of quadrupeds; and they could be used more effectively to hurl stones and sticks at competitors.

Then a whole series of discoveries will have occurred. First, some hominids will have noticed that a stick that smolders or burns on one end is cool enough to be handled on the other. They will have found that such sticks make particularly effective weapons. Entire bands of hominids will have rallied, making noises and brandishing burning sticks to frighten off other animals. Another discovery will have been made subsequently: some individuals will have thrown dry unburning sticks on the flames, and will have made further handy torches for use as a weapon.

The act of igniting the end of a dry stick marked a decisive breakthrough in our species' gamble on intelligence. A natural fire goes out after a time, but one that is kindled with additional sticks keeps burning. Our ancestors will have discovered that by lighting sticks they could not only frighten off other animals but could keep fires going. Since natural fires would not occur at all times—periods without lightning can be long—keeping fires alive became an important chore. And then a third discovery was made: fire could be transported. A burning stick could be carried and made to ignite fires at more convenient locations, for example, in or near caves. Thenceforth fires were laid near human habitations and were used for roasting food as well as for keeping predators at bay. There is evidence that fires were indeed used in this way, and for staggering periods of time. The famous cave at Zhoukoudian, for example, seems to have had a fire that was tended off and on for about 230,000 years—and was abandoned only when the roof collapsed and the cave had to be vacated.

Given further scores of centuries, hominids have discovered that they could make fires on their own, without having to wait for the serendipity of a bolt of lightning igniting dry bush. Rubbing together sticks and stones and blowing on the sparks was a remarkable invention of hominid intelligence. Together with the earlier discoveries, it endowed our ancestors with a significant measure of control over nature, far more than any other creature. With this invention, our species acquired an assured path to dominance. Humans no longer had to struggle for survival in constant fear of more powerful species: they could establish habitations, protect them, and stockpile their staple foods. A Greek myth tells us that Prometheus stole the fire withheld by an avenging Zeus, angry at humans for having had the better of him. The Promethean spark, concealed according to legend in the hollow stalk of a fennel, may have been the greatest breakthrough in the history of sapiens.

With the edge on survival assured, the payoffs of intelligence accumulated at an increasing rate. River valleys, such as the Nile, the Tigris, the Euphrates, the Ganges, and the Huang-Ho were settled. In these environments silt deposited by great streams acted as a natural fertilizer, and periodically flooding waters functioned as natural systems of irrigation. In the course of millennia regular harvests were supplemented by seeds planted on favorable locations; several strains of previously wild plants could be successfully domesticated. The domestication of a few species of animals occurred at more or less the same time. With the advent of the Neolithic Age—a breakthrough that has a rightful claim to be the first great technological "revolution"—the nomadic bands of hominids transformed into settled pasturalists.

The rest is indeed history; the history of sapiens, the dominant predator of this planet. The intelligence we evolved permitted us to reproduce in ever greater numbers and to dominate—or at least to interfere with—nature according to the dictates of our growing needs and our increasingly voracious appetites.

THE QUESTION

We have learned to make fire and have acted upon the assumption that we can also put it out. But is such confidence justified? The forces we have called into being are all fires of one kind or another: dynamic processes in nature that we catalyze and then hope to control. We believe that we have tamed these Promethean fires—that we cannot only create them but can also extinguish them at will. Yet some of the fires we have sparked get out of hand occasionally. Some, like a maverick genie let out of a bottle, take on a life and will of their own. They act in unforeseen and unintended ways, destroying rather than building life and habitat. This was how the force we liberated with the invention of gunpowder behaved, and how most of our fossil-fuel based technologies behave today. As Hiroshima and Chernobyl taught us, the genie we have let out of the nucleus of the atom is more powerful and more difficult to tame than all the others. And robots and computers and the myriad new technologies of automation and communication may not turn out to be reliably domesticated either.

All this should give us food for thought. When the line of Homo branched off from the higher apes some 5 million years ago, our species—and therewith terrestrial nature—took a chance. It put its own continuation at stake. An intelligent species is not necessarily an evolutionary success, reproducing and enhancing its environment. It might also be an ecological disaster, degrading its milieu and threatening its own survival. If human intelligence were to end in a fiasco, the extinction of our species would very likely mean the extinction of all higher forms of life on Earth. The bet on intelligence was the greatest gamble the biosphere has ever made.

Though the outcome was in doubt for millions of years, the bet seems to have paid off for us in the span of recorded history. Yet—could it be that this history is now coming to an end? To envisage the extinction of our species is by no means farfetched—elsewhere in the universe intelligent species may have disappeared not long after they became dominant. Intelligence, after all, is one of the many answers that evolution can offer in the great dance of mutation and natural selection, and it is probable that in the wide reaches of the universe similar answers will have been chanced upon. Despite this, our efforts at interplanetary communication have been a failure. There have been reports of UFO's with extraterrestrials on board landing on Earth, but they are not confirmed and their veracity is seriously questioned. Even though there are many planets capable of supporting life within communication range from Earth, we have not established regular contact with any of them. The reason may be not that intelligent species do not exist beyond our planet, but that, even if a few may be interspersed in the galaxy, such species do not survive for long. If most of them have a short life-expectancy, our chances of communicating with them are drastically reduced. We would have to be precisely coordinated in space and time to receive signals from them: a few hundred years too soon and they would not be capable of emitting the signals; a few hundred years too late and they would no longer be there to emit them.

Whether or it exists elsewhere in the cosmos, we pride ourselves that intelligent life exists here on Earth. But does it? The answer depends on the meaning we attach to "intelligence." As a strategy for competitive survival, intelligence of the human kind does exist: it has paid off handsomely in the last few thousand years. Yet its costs have been rising and now threaten to overtake its benefits. If they do, our species will turn into a planetary parasite that kills the host upon which it feeds—a kind of cancer that destroys the biosphere. And that would hardly be intelligent.

There could be other *kinds* of intelligence; the term means, after all, the ability to make considered choices. In a world that is complex and interdependent, making such choices is not easy. It calls for thinking and acting in a global context, with a long time-horizon. Short-range tunnel vision could prove fatal—it may warrant choices that prove disastrous for the individual who makes them and catastrophic for the biosphere in which his species has evolved.

Will our intelligence test out in the end—will we make the right choices? This is the ultimate question: our collective survival depends on it. When a Stone Age fire got out of hand, a part of the forest or savannah was destroyed and some habitations had to be vacated; the nomadic bands of sapients moved to untouched regions. Throughout the Modern Age, "go West, young man" was a feasible proposition—one could always set out for as yet virgin lands. But today the situation is different. The forces humanity is now unleashing do not leave any region of this planet untouched: if they get out of hand there will not be anywhere left to go. If we make the wrong choices we will use our mega-technologies to dig a mega-grave for ourselves.

THE FARTHERMOST HORIZONS

What if timely shifts in values and motivations could create that subtle but essential fluctuation that would shift our bifurcating societies toward a humanistic evolutionary pathway? What if our evolving systems were impelled into a condition that is both sustainable and humanistic? In other words, what if the "third strategy" of evolutionary humanism were put into practice? The long-term prospects could be breathtaking. Rather than devolving into extinction, sapiens could evolve with distinction.

The next generation of humans could make a major evolutionary leap. Here is an intriguing fact: in terms of human population, we are approaching a magic number: 10^{10}—ten billion. This is how many humans will live on this planet by the time the curve of world population growth finally levels off. The number 10^{10} is closely associated with major evolutionary leaps. As Peter Russell pointed out, it takes some 10 billion atoms to make a basic living cell, and about 10 billion cells to make an autonomous multicellular organism. It also takes 10 billion neurons to create consciousness in the neocortex of the human brain. If life emerges from physical and chemical processes when this threshold is reached, and if consciousness emerges in living beings, significant novelty could also emerge when this many conscious beings congregate within living societies.

Of course, numbers alone provide merely a quantitative parameter and not the full set of conditions to be satisfied if an evolutionary breakthrough is actually to occur. An amalgamous mass of 10 billion atoms could no more create a living cell than a mass of 10 billion cells could create an organism—or a conscious brain. There must be precise connections among the components, cycles within cycles, feedbacks within feedbacks, and coherent integration on the level of the whole. Only than can cellular life emerge in a system of atoms and molecules, and autonomous life and consciousness come about in a system of living cells.

What are the chances that the kinds of cycles and feedbacks that in nature make for a leap to a new evolutionary level would also occur within the human population of this planet? The chances seem rather good. As we have seen, quantitative, extensive growth is now leveling off and is likely to give way to qualitative, intensive development, that is, to structuralization and complexification. After all, not only more people, energy, and matter, but ever more information is injected into our social systems, and information always structures, and not merely agglomerates, in the system into which it flows. If this process were to continue, the kind of developmental rhythm that is typical of the growth of the embryo in the womb would be re-played on the level of entire human populations.

The fact is that the genesis of the brain in the embryo is remarkable both for its precision, and for the analogies it offers for the likely growth of the world population. The growth of brain cells accelerates from about the eighth week of gestation; by the tenth week it becomes explosive. A million cells are added to the fetal brain each and every minute. Then, at the thirteenth week, extensive growth stops and development turns inward. Instead of growing in numbers, the embryo's brain grows in connections. In a matter of months, the complex structure of the spaiens brain, the product of some 50 million years of evolution, is precisely reconstructed.

The possibility that a similar process could take place on the level of human populations cannot be lightly dismissed. Contemporary people are already organized into complex structures in cities and villages, public institutions and private enterprises, professional guilds and associations, social clubs and cultural bodies, and a myriad of other groups. These structures are connected through multiple cycles and feedbacks, and their interconnections continue to grow at an exponential rate. Writing is 10,000 years old, and printing about 500. The telegraph and the telephone are products of the nineteenth century, and radio communication appeared in the twentieth century. The widespread use of computers for information and communication dates only from the 1960 and telex and telefax are still more recent developments. Presently, the number of computer networks is doubling every few years; and radio, television, telex and fax penetrate every corner of the planet. The world population is becoming both extensively and intensively interconnected.

Seeing that connections are rapidly growing among a critical mass of natural components, the question arises whether or not a new evolutionary phenomenon may be in the offing. Could the outcome be something like a world superbrain in which human individuals are mere information-transmitting neurons? This possibility—highlighted in a fascinating but frightening vein in Peter Russell's concept of the "global brain," and in a more spiritual form in Teilhard de Chardin's idea of noospheric evolution—is not the only option. It would come about only if we allowed the cycles and feedbacks that interconnect us to evolve at the expense of our individual freedom and autonomy. This would amount to a hierarchic evolutionary scenario. Fortunately, we are evolved enough, and out brains are conscious enough, to steer away from enslaving scenarios and opt for a holarchic path where individuals and communities collaborate of their own accord in flexible social systems. We are, after all, entering the Aquarian Age, participatory and humane, yet disciplined—and lasting, so the astrologers tell us—for over two thousand years.

The far horizons of the future underscore the alternatives that now open for humanity. Devolution into extinction, or evolution to a new plateau of existence; the stakes have always been great but never as great as today. The responsibilities are likewise the greatest they have ever been. Having become conscious of evolution, we must now make evolution itself conscious. If we so willed it, the next leap in the development of human society can be intentionally guided. In the ultimate decade of this century we have the means, and the opportunity, to design our destiny.

APPENDIX

Basic Concepts of Evolutionary Systems Theory

The new sciences of evolving systems trace their origins to the general system theory of Ludwig von Bertalanffy, the cybernetics of Norbert Wiener, and the information theory of Claude Shannon. The basic concepts and theories were developed in various domains of the natural and social sciences as well as in philosophy. They achieved maturity with Ilya Prigogine's nonequilibrium thermodynamics and current advances in the mathematical modeling of chaos and transformation in dynamical systems.

These sciences give us a fresh view of the nature of reality. In this view man and society are not strangers in the universe but integral parts of the great sweep of evolution that began with the Big Bang 18 billion years ago and now issues in the phenomena of life, culture, and consciousness. The new sciences describe the dynamical features of this evolution and its major stages. If there is a sound basis for assessing the next step in the evolution of humankind, and for attempting to steer it in the joint interest, surely these sciences are in a privileged position to provide it. Familiarity with their key conceptions is part of the basic literacy of our time.*

THE FOUNDATIONS

Matter in the universe configures into more and more complex entities where the parts cohere together and share the same destiny. These configured entities are known as systems. Not all systems in the world are the same, although there are general categories that cut across the traditional divisions of

*For further details see: Ervin Laszlo, *Evolution: The Grand Synthesis*, Boston and London: New Science Library, Shambhala Publications, 1987.

the natural and the social sciences. The new categories refer not to "physical system," "chemical system," "biological system," etc., but to states at, near to, or far from, equilibrium. Systems that are far from equilibrium have not been known for long; yet they make up the category of systems that evolve in the physical, as well as in the biological and the human world. The other two categories of systems have been known for over a century.

In equilibrium systems the flows of energy and matter have eliminated differences in temperature and concentration; the elements of the system are unordered in a random mix and the system itself is homogeneous and dynamically inert. In systems near (but not in) equilibrium there are small differences in temperature and concentration; the internal structure is not random and the system is not inert. Such systems tend to move toward equilibrium as soon as the constraints which keep them in nonequilibrium are removed. Systems in this state reach equilibrium when the forward and reverse reactions compensate one another statistically, so that there is no longer any overall variation in the concentrations (a result known as the law of mass action, or Guldberg and Waage's law). The elimination of differences between concentrations means chemical equilibrium, and reaching a uniform temperature means thermal equilibrium. While in a state of nonequilibrium the systems perform work and therefore produce entropy, at equilibrium no further work is performed and entropy production ceases. In a condition of equilibrium the production of entropy, and forces and fluxes (the rates of irreversible processes) are all at zero, while in states near equilibrium entropy production is small, the forces are weak and the fluxes are linear functions of the forces. Thus a state near equilibrium is one of linear nonequilibrium, described by linear thermodynamics in terms of the statistically predictable tendency toward the maximum dissipation of free energy and the highest level of entropy. Systems in the second state will ultimately reach a state characterized by the least free energy and the maximum entropy compatible with boundary conditions, whatever the initial conditions.

The third possible category is that in which systems are far from thermal and chemical equilibrium. Such systems are nonlinear and pass through indeterminate phases. They do not tend toward minimum free energy and maximum specific entropy but amplify certain fluctuations and evolve toward a new dynamic regime that is radically different from stationary states at or near equilibrium.

Prima facie the evolution of systems in the far-from-equilibrium state appears to contradict the famous Second Law of Thermodynamics. How can systems actually increase their level of complexity and organization, and become more energetic? The Second Law states that in any isolated system organization and structure tend to disappear, to be replaced by uniformity and randomness. Contemporary scientists know that evolving systems are not isolated, and thus that the Second Law does not fully describe what takes place in them—more precisely, between them and their environment. Systems in the third category are always and necessarily open systems, so that change of entropy within them is not determined uniquely by irreversible internal processes. Internal processes within them do obey the Second Law: free energy, once expanded, is unavailable to perform further work. But energy available to perform further work can be "imported" by open systems from their environment: there can be a transport of free energy—or negative entropy—across the system boundaries.* When the two quantities—the free energy within the system, and the free energy transported across the system boundaries from the environment—balance and offset each other, the system is in a steady (i.e., in a stationary) state. As in a dynamic environment the two terms seldom balance each other for any extended period of time, in the real world systems are at best "metastable": they tend to fluctuate around the states that define their steady states, rather than settle into them without further variation.

These basic concepts are applied, tested and elaborated in a number of scientific fields and in a variety of ways. Research directly concerned with evolutionary concepts can be roughly divided into two categories: empirical research relying on observation and experimentation, and theoretical research on formal—mathematical and topological—models of systems behavior.

---

*Change in the entropy of the systems is defined by the well-known Prigogine equation $dS = d_iS + d_eS$. Here dS is the total change of entropy in the system, while d_iS is the entropy changed produced by irreversible processes within it and d_eS is the entropy transported across the system boundaries. In an isolated system dS is always positive, for it is uniquely determined by d_iS, which necessarily grows as the system performs work. However, in an open system d_eS can offset the entropy produced within the system and may even exceed it. Thus dS in an open system need not be positive: it can be zero or negative. The open system can be in a stationary state ($dS = 0$), or it can grow and complexity ($dS \leq 0$). Entropy change in such a system is given by the equation $d_eS - d_iS \leq 0$); that is, the entropy produced by irreversible processes within the system is shifted into the environment.

EMPIRICAL RESEARCH

The starting point for empirical research is the observed fact that, under suitable conditions, a constant and rich energy flow passing through a system drives it toward states characterized by a higher level of free energy and a lower level of entropy. As Ilya Prigogine predicted in the 1960s, and as experiments performed by biologist Harold Morowitz confirmed already in 1968, a flow of energy passing through a nonequilibrium system in the third state organizes its structures and components and enables it to access, use and store increasing quantities of free energy. At the same time, the complexity of the system increases and its specific entropy decreases.

The significant measure in evolution is not the gross increase in free energy in the system, but the increase in the density of the free energy flux that is accessed, retained and then used in it. "Energy flux density" is a measure of the free energy per unit of time per unit of volume; for example, erg/second/cm^3. As we ascend the scale of complexity in systems we find that the amount of free energy flux density (i.e., the amount of free energy per time per volume in the system) increases. A complex chemical system retains more of this factor than a monatomic gas; a living system retains more than a complex chemical system. This indicates a basic direction in evolution, an overarching sweep that defines the arrow of time in the physical as well as in the living universe.

The relationship between energy flow over time and change in specific entropy and free energy flux density is essential for answering not only the question as to *how* systems in the third state evolve, but also whether they evolve *necessarily* when certain conditions are present. Until the 1970s, investigators leaned toward the view-exposed eloquently by the French physicist Jacques Monod, that evolution is due mainly to accidental factors. But as of the 1980s many scientists have become convinced that evolution is not an accident, but occurs necessarily whenever certain parametric requirements have been fulfilled.

Laboratory experiments and quantitative formulations corroborate the non-accidental character of the evolution of third-state systems. Ordered structure always emerges when complex systems are immersed in a rich and enduring energy flow. The principles responsible for this evolutionary phenomenon are the following. First of all the system must be open, that is, it must be fed initial reactants and allowed to discharge its final products. Then, the system must have sufficient diversity of components and complexity of structure to be stable in more than one dynamical steady state (i.e., it must have multistability). Last but not least, there must be feedbacks and catalytic cycles among the system's principal components.

The requirement for catalytic cycles has solid roots. In the course of time such cycles tend to be naturally selected in virtue of their remarkable capacity for persistence under a wide range of conditions. Catalytic cycles have great stability and produce fast reaction rates. They come in two flavors: auto-catalysis, where a product of a reaction catalyzes its own synthesis, and cross-catalysis, where two different products (or groups of products) catalyze each other's synthesis.*

In relatively simple chemical systems autocatalytic reactions tend to dominate, while in more complex processes, characteristic of living phenomena, entire chains of cross-catalytic "hyper" cycles appear. For example, as physicist Manfred Eigen has shown, nucleic acid molecules carry the information needed to reproduce themselves as well as to produce an enzyme. The enzyme catalyzes the production of another nucleic acid molecule, which in turn reproduces itself and produces another enzyme. The loop may involve a large number of elements; ultimately it closes in on itself, forming a cross-catalytic reaction cycle remarkable for its fast reaction rates and stability under diverse parametric conditions. It is not surprising that catalytic hypercycles underlie the stability of the sequence of nucleic acids that code the structure of living organisms; and that they also underlie, on higher evo-

---

*An example of auto-catalysis is the reactions scheme $X + Y \rightarrow 2X$. Starting from one molecule of X and one of Y, two molecules of X are catalyzed. The chemical rate equation for this reaction is $dX/dt = k\,X\,Y$. When Y is held at a constant concentration (k), there is an exponential growth in X. Cross-catalytic reaction cycles have been studied in detail by the Brussels school of Ilya Prigogine. A model of such reactions, known as the Brusselator, consists of the following four steps:

(1) $A \rightarrow X$

(2) $B + X \rightarrow Y + D$

(3) $2X + Y \rightarrow 3X$

(4) $X \rightarrow E$

In this reaction model X and Y are intermediate molecules within an overall sequence through which A and B become D and E. In step (2) Y is produced from X, while in step (3) an additional X is produced through collisions of 2X and Y. Thus while (3) in itself constitutes auto-catalysis, (2) and (3) in combination make for cross-catalysis.

lutionary levels, the persistence of organic species as well as of entire ecologies in our planet's biosphere.

Given sufficient time, and an enduring energy flow acting on organized systems within permissible parameters of intensity, temperature and concentration, the basic catalytic cycles tend to interlock within emerging hypercycles. This process evolutionary systems theory identifies as *convergence.* Convergence does not lead to growing similarities among systems and ultimately to uniformity (as in the convergence of ideologies and socioeconomic systems), since the evolving systems functionally complete and complement each other. Through the process of evolutionary convergence new, higher level systems are created that selectively disregard many details of the dynamics of their subsystems and impose an internal constraint that forces the subsystems into a collective mode of functioning. This mode, which is that of the emerging systems themselves, is simpler than the sum of the uncoordinated functions of the subsystems.

Convergence occurs in all realms of evolution. It is in virtue of creating progressively higher level systems with an initially simpler structure that evolution can unfold. On each level third-state systems exploit the free energy fluxes in their environment. As the density of the free energy retained in the systems increases, the systems acquire structural complexity. If the process would continue indefinitely, a functional optimum would be reached, beyond which further increases in complexity would no longer contribute to dynamical efficiency; thereafter evolution could only produce nonselective drift. However, due to the convergence of third-state systems on successively higher levels of organization, structurally simpler suprasystems recapitulate the process whereby free energy densities are increasingly exploited through structures of growing complexity.

In sum, the processes of evolution create initially comparatively simple dynamical systems on particular levels of organization. The processes then lead to the progressive complexification of the existing systems and, ultimately, to the creation of simpler systems on the next higher organizational level, where complexification begins anew. Thus evolution moves from the simpler to the more complex, and from the lower to the higher level of organization.

The empirical evidence for this process is indisputable. Diverse atomic elements converge in molecular aggregates; specific molecules converge in crystals and organic macromolecules; the latter converge in cells and the subcellular building blocks of life; single-celled organisms converge in multicellular species; and species of the widest variety converge in ecologies. As each level of organization is attained, increasingly complex systems evolve on that level. On the level of atoms, structures build in time from hydrogen to uranium and beyond; on that of molecules simple chemical molecules are followed by the synthesis of more complex polymers; on the organic level species evolve from unicellular to multicellular forms, and on the still vaster ecological level immature ecosystems build toward the mature climactic form.

Change occurs in systems and evolution unfolds because dynamical systems in the third state are not stable. They have upper thresholds of dynamical stability which, in a changing environment, tend to be transgressed. When they are, critical instabilities occur in the systems. Experiments show that dynamical systems far from equilibrium can be "kicked out" of their steady states by changes in the critical parameters. The systems prove to be highly sensitive to changes in those parametric values that define the functioning of their catalytic cycles. When these values change, the systems enter a transitory phase characterized by indeterminacy and chaos, and a sudden increase in entropy production. The transitory phase comes to an end when the systems disorganize into their stable subsystems—or find a new set of dynamic steady states. If they do not vanish as complex entities in their own right, the systems evolve a new dynamic regime. In this regime they are again maintained by catalytic cycles and multiple feedbacks, and their entropy production drops to a functional minimum.

The way in which dynamical systems respond to destabilizing changes in their environment is of the greatest importance for understanding the dynamics of evolution in the diverse realms of nature. Dynamical systems do not evolve smoothly and continuously over time, but do so in comparatively sudden leaps and bursts. Real-world systems can evolve through sequences of destabilizations and phases of indeterminacy since they have multiple states of stability—when one steady state is fatally disrupted others remain accessible. The further the systems are from thermodynamic equilibrium, the more sensitive is their structure to change and the more sophisticated are the feedbacks and catalytic cycles that maintain them.

According to current scientific conceptions, the selection from among the set of dynamically functional alternative steady states is not predetermined. It is due neither to initial conditions in the system nor to manipulations of the critical parametric values. At the critical junctures, when they are critically destabilized and in chaos, complex systems act indeterminately: one among their possibly numerous internal fluctuations is amplified, and the amplified fluctuation spreads with great rapidity within the system. The amplified or "nucleated" fluctuation dominates the system's new dynamic regime and determines its new steady state.

ADVANCES IN THEORY

The observed dynamics of the evolution of complex systems call for the development of new theoretical tools. This is true especially in regard to the discontinuous, nonlinear nature of change in dynamic systems which differential calculus, the mathematics used conventionally to model change, is poorly equipped to handle. In its standard form, differential calculus assumes change to be smooth and continuous.

The contemporary offshoot of classical dynamics, dynamical systems theory, has risen to the challenge. Dynamical systems theorists elaborate mathematical models of complex system behavior not only for the intrinsic theoretical interest of the models, but also for their possible application to complex systems in the empirical world. The models (consisting of ordinary differential equations, partial differential equations of the evolution type and finite difference equations singly or in sets) simulate the dynamical aspects of the behavior of complex systems. The development of the simulation models is not limited to the range of their actual application: dynamical systems theorists explore all possible models within range of their mathematical tools and then search for varieties of empirical systems to which the models may apply. This hypothetico-deductive approach generates a wide variety of models and simulations and promises to significantly enhance our understanding of discontinuous transformations in the behavior of many varieties of complex systems.

In the language of dynamical systems theory, static, periodic, and chaotic attractors govern the long-run behavior of complex systems. A static attractor "traps" the trajectory of the states of a system—its time series—so that the system comes to rest at a stable state. A periodic attractor traps the trajectory in a cycle of states that repeats in a given time interval; the system is then in an oscillatory state. The chaotic attractor, in turn, provokes a quasi-random, chaotic series of states, with the system neither coming to rest nor settling into an oscillatory mode but continuing to exhibit erratic, but by no means disordered, behavior.

In recent years chaotic behavior has been discovered in a wide variety of systems. Such behavior is exhibited by processes as varied as fluids in flow, and the blending of substances during solidification. The phenomenon of turbulence is a case in point: it has been known since the nineteenth century, but its origins have been imperfectly understood. By 1923, experiments in fluid dynamics had demonstrated the appearance of annular Taylor vortices; these appear when the speed of stirring in a fluid increases beyond a critical point. Further increases in stirring produce additional abrupt transformations and ultimately turbulence. And turbulence is a paradigm for the chaotic state.

The behavior of complex systems in the empirical world is normally influenced by many different attractors simultaneously; dynamical systems theory accounts for such systems by correspondingly complex models. In the models major and abrupt changes in system behavior represent bifurcations. These appear in the phase portrait of the systems and are due to changes in the "controls" that make up the critical parameters. Bifurcations are modeled as a shift from one type of attractor to another, for example, from a static to a periodic attractor. A hitherto stable system begins to oscillate, or, in a shift from periodic to chaotic attractors, a hitherto oscillating system lapses into chaos. These so-called "subtle" bifurcations are but one variety of basic system changes; there are also "explosive" and "catastrophic" bifurcations. Catastrophic bifurcations (which do not mean catastrophes in the everyday meaning of the term) consist of the sudden appearance of disappearance "out of the blue" of static, periodic, or chaotic attractors.

The bifurcations mapped by dynamical systems theorists have significant application in real-world systems. Subtle bifurcations represent increasing instability in systems far from thermodynamical equilibrium. A system, such as a series of chemical reactions in stable equilibrium, begins to oscillate; or an oscillating system, such as a chemical clock, becomes turbulent. In its mathematical models dynamical systems theory identifies a number of such "scenarios" leading from stable equilibrium to chaos. Models with catastrophic bifurcations leading from turbulent to newly ordered states through the reconfiguration of the attractors simulate evolutionary processes in real-world systems in the third state. Bifurcations are the kind of transformations that underlie the evolution of all varieties of real systems, from the atoms of the elements to organic species and entire ecologies and societies.

IN CONCLUSION

The thrust of contemporary approaches to the evolution of complex systems can be briefly summarized. The basic elements are nonequilibrium systems maintained by catalytic cycles within enduring flows of energy; the alternation of determinate order in periods of stability with states of creative chaos during bifurcations; and the observed statistical tendency toward greater complexity on sequentially higher levels of organization.

Auto-catalytic and cross-catalytic feedback loops predominate in open dynamical systems far from equilibrium in virtue of their rapid reaction rates and great stability. However, as no self-stabilizing reaction cycle is entirely immune to disruption, constant changes in the environment sooner or later produce conditions under which certain self-stabilizing cycles can no longer operate. The systems encounter a point known in dynamical systems theory as catastrophic bifurcation. The outcome in third-state systems, as both experiment and theory demonstrate, is essentially indeterminate: it is not a function of initial conditions, nor of changes in the control parameters. There is an observed probability that bifurcations lead to increasingly complex systems progressively further from thermodynamic equilibrium. In the course of time the systems retain a more dense flux of free energy for a longer period and decrease their specific entropy. Without this probability evolution would produce a random drift between more and less organized states, instead of a statistically irreversible build-up of increasingly complex and dynamic nonequilibrium systems.

Progressively higher levels of organization are attained as catalytic cycles on one level interlock and form hypercycles: these are systems on a higher level. Thus molecules emerge from combinations of chemically active atoms; protocells emerge from sequences of complex molecules; eukaryotic cells emerge among the prokaryotes; metazoa make their appearance among the protozoa and converge in still higher-level ecological and social systems.

These factors and processes hold true in all domains of nature, from the most basic level of particles and atoms swirling in the infinite reaches of the universe to the most complex level of organisms forming ecologies and societies within the Earth's biosphere.

FURTHER READING
Classics of Evolutionary Systems Theory

(Compiled by Alexander Laszlo)

Ackoff, Russell L., General Systems Theory and systems research: Contrasting conceptions of systems science. In *Views on a General System Theory: Proceedings from the Second System Symposium*. M. Mesarovic, ed. New York: John Wiley & Sons, 1964.

Angyal, A., *Foundations for a Science of Personality*. Cambridge: Harvard University Press, 1961.

Ashby, W. Ross, Principles of the Self-organizing system. In *Principles of Self-Organization*. Foerster and Zopf, eds. New York: Pergamon Press, 1962.

___, *An Introduction to Cybernetics*. London: Chapman & Hall; New York: Barnes & Noble, 1956.

Battista, John R., The holistic paradigm and General system Theory. In *General Systems*. 1977, **22**:65-71.

Bennis, Warren, et al., eds., *The Planning of Change*. New York: Holt, Rinehart & Winston, 1962.

Beer, Stafford, *Platforms of Change*. New York: John Wiley & Sons, 1979.

Beishon, J., and G. Peters, *Systems Behavior*. New York: Open University Press, 1972.

Berger, Peter L., and Thomas Luckman, *The Social Construction of Reality: A Treatise in the Sociology of Knowledge*. New York: Doubleday, 1966.

Bertalanffy, Ludwig von, *Perspectives on General System Theory: Scientific-Philosophical Studies*. Edgar Taschdjian, ed. New York: George Braziller, 1975.

___, *General System Theory: Essays on its Foundation and Development* [rev. ed.], New York: George Braziller, 1968.

___, *Robots, Men and Minds*. New York: George Braziller, 1967.

___, General System Theory—a critical review. In *General Systems*. 1962, **7**:1-20.

___, *Problems of Life: An Evaluation of Modern Biological Thought*. New York: John Wiley & Sons, 1952.

Bloomfield, Brian P., *Modelling the World: The Social Construction of Systems Analysts*. New York: Basil Blackwell, 1986.

Blum, Harold F., *Time's Arrow and Evolution* [3rd edition], Princeton, N.J.: Princeton University Press, 1968.

Boulding, Kenneth E., *Ecodymanics: A New Theory of Societal Evolution*. London: Sage Publications, 1981.

__, *The Image: Knowledge in Life and Society*. Michigan: Ann Arbor Paperbacks, 1961.

__, General Systems Theory—the skeleton of science. In *Management Science*. 1956, **2**:197-208.

__, *The Organizational Revolution: A Study in the Ethics of Economic Organization*. New York: Harper, 1953.

Bowler, T. Downing, *General Systems Thinking: Its Scope and Applicability*. New York: Elsevier North Holland, 1981.

Buckley, Walter, ed., *Modern Systems Research for the Behavioral Scientist*. Chicago: Aldine, 1968.

__, *Sociology and Modern Systems Theory*. Englewood Cliffs, N.J.: Prentice-Hall, 1967.

Cavallo, Roger E., ed., *Systems Research Movement: Characteristics, Accomplishments, and Current Developments*. Louisville, KY: Society for General Systems Research, 1979.

Checkland, Peter, *Systems Thinking, Systems Practice*. New York: John Wiley, 1981.

Churchman, C. West, *The Systems Approach and Its Enemies*. New York: Basic Books, 1979.

__, *The Systems Approach* [rev. and updated.], New York: Harper & Row, 1979.

__, *The Design of Inquiring Systems: Basic Concepts of Systems and Organizations*. New York: Basic Books, 1971.

Churchman, C. West and R. L. Ackoff, *Methods of Inquiry: An Introduction to Philosophy and Scientific Method*. St Louis: Educational Publications, 1950.

Davidson, Mark, *Uncommon Sense: The Life and Thought of Ludwig von Bertalanffy (1901–1972), Father of General Systems Theory*. Foreword by R. Buckminster Fuller, and introduction by Kenneth E. Boulding. Los Angeles: J. P. Tarcher, 1983.

Demerath, N. J., and R. A. Peterson, eds., *System, Change and Conflict*. New York: Free Press, 1967.

Deutsch, Karl W., *The Nerves of Government*. New York: Free Press, 1963.

Dobzhansky, Theodosius, *The Biology of Ultimate Concern*. New York: New American Library, 1967.

Easton, David, *A Systems Analysis of Political Life*. New York: John Wiley, 1965.

Elsasser, Walter M., *Atom and Organism: A New Approach to Theoretical Biology*. Princeton, N.J.: Princeton University Press, 1966.

Emery, F. E., ed., *Systems Thinking: Selected Readings*. England: Penguin Books, 1969.

Errington, Paul L., *Of Predation and Life*. Ames, Iowa: Iowa State University Press, 1967.

Falk, Richard, Samual S. Kim, and Saul H. Mendlovitz, eds., *Toward a Just World Order*. Boulder, Colorado: Westview Press, 1982.

Foerster, Heinz von, and George W. Zopf, Jr., eds., *Principles of Self-Organization: University of Illinois Symposium on Self-Organization, 1961*. New York: Pergamon Press, 1962.

Fuller, Buckminster, *Operating Manual for Spaceship Earth*. Carbondale, Ill.: Southern Illinois University Press, 1970.

Gerard, W. E., Hierarchy, entitation, and levels. In *Hierarchical Structures*. Whyte, Wilson, and Wilson, eds. New York, 1969.

Gharajedaghi, Jamshid, *Toward a Systems Theory of Organization*. California, Intersystems Publications, 1985.

Gray, William, and N. D. Rizzo, eds., *Unity Through Diversity: A Festschrift for Ludwig von Bertalanffy*. New York: Gordon and Breach Science Publishers, 1973.

Gray, W., F. D. Duhl, and N. D. Rizzo, eds., *General Systems Theory and Psychiatry*. Boston: Little, Brown & Company, 1969.

Grinker, Roy R., ed., *Toward a Unified Theory of Human Behavior*. New York: Basic Books, 1956.

Hall, Edward T., *The Hidden Dimension*. New York: Doubleday, *1966*.

Harris, Dale, ed., *The Concept of Development: An Issue in the Study of Human Behavior*. Minneapolis, Minn.: University of Minnesota Press, 1957.

Harris, Errol E., *The Foundations of Metaphysics in Science*. London: George Allen and Unwin, 1965.

Huxley, Julian, *Evolution in Action*. New York: Harper, 1953.

Katsenelinboigen, Aron, *Some New Trends in System Theory*. Seaside, Calif.: Intersystems Publications, 1984.

Klir, George J., ed., *Trends in General Systems Theory*. New York: Wiley-Interscience, 1972.

Knorr, Klaus, and Sidney Verba, eds., *The International System: Theoretical Essays*. Princeton, N.J.: Princeton University Press, 1961.

Koestler, Arthur, and J. R. Smythies, eds., *Beyond Reductionism: New Perspectives in the Life Sciences*. London and New York: Macmillan, 1969.

Kroeber, Alfred L., *Configurations of Culture Growth*. Berkeley and Los Angeles, Calif.: University of California Press, 1944.

Kuhn, Alfred, *The Study of Society: A Unified Approach*. Homewood, Ill.: Irwin, 1963.

Lange, Oskar, *Wholes and Parts: A General Theory of System Behavior*. New York: Pergamon Press, 1956.

Laszlo, Ervin, The meaning and significance of General System Theory. In *Behavioral Science* **20**, no. 1 (January, 1975):9-24.

__, *A Strategy for the Future: The Systems Approach to World Order*. New York: George Braziller, 1974.

__, *Evolution: The Grand Synthesis*. Boston and London: New Science Library, Shambhala Publications, 1987.

__, *Introduction to Systems Philosophy: Toward a New Paradigm of Contemporary Thought*. New York: Gordon & Breach Science Publishers, 1972; 1984; New York: Harper & Row, 1973.

__, *System, Structure and Experience*. New York: Gordon and Breach, 1969.

__, *Systems Science and World Order: Selected Studies*. Oxford: Pergamon Press, 1983.

__, ed., *The Relevance of General Systems Theory: Papers Presented to Ludwig von Bertalanffy on His Seventieth Birthday*. New York: George Braziller, 1972.

__, ed., *The World system: Models, Norms, Applications*. New York: George Braziller, 1974.

Lerner, Daniel S., ed., *Parts and Wholes*. New York: Free Press, 1963.

Lewin, Kurt, *Field Theory in Social Science: Selected Theoretical Papers*. Dorwin Cartwright, ed. New York: Harper, 1951.

Lilienfeld, Robert, *The Rise of Systems Theory: An Ideological Analysis*. New York: Wiley, 1978.

McCulloch, Warren S., *Embodiments of Mind*. Cambridge, Mass.: MIT Press, 1965.

Margenau, Henry, ed., *Integrative Principles of Modern Thought*. New York: Gordon and Breach, 1972.

__, *Open Vistas: Philosophical Perspectives of Modern Science*. New Haven, Conn.: Yale University Press, 1961.

Maslow, Abraham H., ed., *The Psychology of Science*. New York: Harper & Row, 1966.

__, *New Knowledge in Human Values*. New York: Harper, 1959.

__, ed., *Motivation and Personality*. New York: Harper, 1954: new edition, 1970.

Mather, Kirtley F., *The Permissive Universe*. New York: Gordon & Breach, 1974.

Matson, Floyd W., *The Broken Image*. New York: Braziller, 1964.

Menninger, K., M. Mayman, and P. Pruyser, *The Vital Balance: The Life Process in Mental Health and Illness*. New York: Viking Press, 1963.

Miller, James G., Living systems: basic concepts. In *General Systems Theory and Psychiatry*. William Grey, D. F. Duhl, and N. Rizzo, eds. Boston, 1969.

Morgan, Lloyd, *Emergent Evolution*. London: Williams and Norgate, 1923.

Nappelbaum, E. L., Yu A. Yaroshevskii, and D. G. Zaydin, *Systems Research: Methodological Problems*. (Prepared by USSR State Committee for Science and Technology, USSR Academy of Sciences, Institute for Systems Studies.) Oxford and New York: Pergamon Press, 1984.

Northrop, F. S. C., *The Logic of the Sciences and the Humanities*. New York: Macmillan, 1947.

Parsons, Talcott, *Structure and Process in Modern Societies*. Glencoe, Ill.: Free Press, 1960.

___, *The Social System*. New York: Free Press, 1957.

Parsons, Talcott, E. A. Shils, K. D. Naegele, and T. R. Pitts, eds., *Theories of Society*. New York: Free Press, 1961.

Portmann, Adolf, *Animals As Social Beings*. New York: Viking Press, 1961.

Rapoport, Anatol, *General System Theory: Essential Concepts and Applications*. Cambridge, Mass.: Abacus Press, 1986.

___, General System Theory. In *The International Encyclopedia of Social Sciences*. David L. Sills, ed. New York: The Macmillan Publishing Co. & The Free Press, 1968.

Rosenau, James N., ed., *Linkage Politics*. New York: Free Press, 1969.

Rosenblueth, Arturo, *Mind and Brain: A Philosophy of Science*. Cambridge, Mass.: MIT Press, 1970.

Russell, Peter, *The Awakening Earth: The Global Brain*. London: Routledge, 1982.

Schrödinger, Erwin, *What is Life?* Cambridge: Cambridge University Press, 1945.

Selye, Hans, *The Stress of Life*. Toronto and New York: McGraw-Hill, 1956.

Simon, Herbert A., *The Sciences of the Artificial*. Cambridge, Mass.: MIT Press, 1969.

Sinnot, Edmund W., *The Problem of Organic Form*. New Haven, Conn.: Yale University Press, 1963.

Sorokin, Pitirim A., *Sociological Theories of Today*. New York: Harper & Row, 1966.

Stanley-Jones, D., and K. Stanley-Jones, *The Kybernetics of Natural Systems: A Study in Patterns of Control*. New York: Pergamon, 1960.

Teilhard de Chardin, Pierre, *The Phenomenon of Man*. New York: Harper & Row, 1965.

Thayer, Lee, ed., *Communication: General Semantics Perspectives*. New York: Spartan, 1970.

Varela, Francisco J., Autonomy and autopoiesis. In *Self-Organizing Systems: An Interdisciplinary Approach*. Gerhard Roth and Helmut Schwegler, eds. Frankfurt: Campus Verlag, 1981.

Waddington, C. H., ed., *Towards a Theoretical Biology*. Chicago: Aldine, 1970.

Walter, W. Grey, The past and future of cybernetics in human development. In *Progress of Cybernetics*. J. Rose, ed. 1970, **1**:45-56.

___, *The Living Brain*. London and New York: Norton, 1953.

Weiss, Paul A., ed., *Hierarchically Organized Systems in Theory and Practice*. New York: Hafner Publishing Co., 1971.

__, *Dynamics of Development: Experiments and Inferences*. New York: Academic Press, 1968.

Whitehead, Alfred North, *Science and the Modern World*. New York: Macmillan, 1925.

__, *The Concept of Nature*. Cambridge: Cambridge University Press, 1920.

Whorf, Benjamin Lee, *Language, Thought and Reality: Selected Writings of B. L. Whorf*. John B. Carroll, ed. New York: Wiley, 1956.

Whyte, Lancelot Law, *The Next Development in Man*. New York: Mentor Books, 1950.

__, *Unitary Principles in Physics and Biology*. New York: Henry Holt, 1949.

Whyte, L. L., A. G. Wilson, and D. Wilson, eds., *Hierarchical Structures*. New York: American Elsevier, 1969.

Wiener, Norbert, *The Human Use of Human Beings: Cybernetics and Society* [2nd ed.], Garden City, N.Y.: Doubleday Anchor Books, 1954.

Woodger, Joseph H., *Biological Principles*. New York: Humanities, 1966.

__, *Biology and Language*. Cambridge: Cambridge University Press, 1952.

SUBJECT AND NAME INDEX